Your Guide to Cost Reduction through Pneumatics Automation

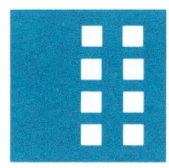

- **You Already Are a Master of Pneumatics Automation** 2
  What it takes to automate with pneumatics.

- **How to Identify Cost Reduction Opportunities** 3
  What to look for when targeting potential cost reductions.

- **Pneumatics Automation as a Cost Reduction Tool** 4-5
  The role of pneumatics in the cost reduction process.

- **Pneumatics Automation at Work** 6-17
  A look at how pneumatics can automate operations. Examples include clamping, facing, assembly, drilling, broaching, and loading.

- **How to Carry Out Your Own Pneumatics Automation Project** 18-19
  A step-by-step guide to implementing simple automation through pneumatics.

- **A Pneumatics Automation Trouble-Shooting Guide** 20-21
  What to check first when things go wrong.

- **Pneumatic Components** 22-53
  A review of some of the many pneumatic components that will build productivity as they cut costs.

- **Where to Learn More about Pneumatics** 54-55
  A selected listing of materials on pneumatic components and applications.

- **Sources for Pneumatic Components** 56-63
  A member list of the National Fluid Power Association's pneumatics manufacturers and the products they offer.

- **Index** 64

You Already Are a Master of Pneumatics Automation

Have you ever...

- Designed or built a fixture, mold or template?
- Done major repairs on a pneumatic, hydraulic or electrical piece of assembly, machining or conveying equipment?
- Set up a manual assembly line with powered presses, fastener insertion or automated testing?
- Designed a product and had a part in setting up its production processes?

The above tasks contain most of the skills needed to successfully complete a simple pneumatics automation project. Simple pneumatics automation differs from large-scale, integrated automation even though it utilizes many of the same components and techniques.

Integrated automation projects often involve centrally-controlled, multi-station manufacturing processes. These complex assembly projects can cost hundreds of thousands of dollars and take months or years to complete. Simple pneumatics automation projects can be completed with far fewer resources expended since they are usually focused upon mechanizing a single task. If necessary, they can later be incorporated into an assembly of machines.

Simple pneumatics automation harnesses the power of today's low-cost, understandable electronic, electrical and mechanical technologies. It utilizes standard components such as actuators, valves, sensors, programmable controllers, grippers, air motors, and other similar devices which are designed to be readily assembled into a working automated application.

Of course as you automate with pneumatics, be sure to read and observe all use and safety related instructions and recommendations of the manufacturer and supplier involved. Equipment should be installed and used in accordance with the manufacturers' and suppliers' instructions. Take the time to properly train your employees in the correct and safe method of operation of automated equipment. A safe work environment is essential for employee well-being and satisfaction, and makes good business sense.

How to Identify Cost Reduction Opportunities

Reducing costs can be easy when you take the process in steps, working on one project at a time. Where do you start? Here are some clues. Take a walk through your plant and look for these opportunities to save yourself some time and money:

- Look for idle hands. If a machine operator is forced to stand idly, watching an automatic machine work, you are wasting his time and your money.

- Look for awkward or time-consuming loading and unloading operations. Loading and unloading are among the easiest work functions to make more efficient through automation — without a lot of expense.

- Look for conditions where a single operator could run several similar machines at once.

- Look for situations where a machine performs a single operation when several could be performed at once. Consider attacking the workpiece from several directions at the same time.

- Look for hazardous work areas where automatic feeding could replace hand feeding of parts, or where the machine trip mechanism should be altered to require safer, two-hand actuation.

- Look for opportunities to use automated fixtures which improve operational efficiency.

An examination of your cost-accounting data will reveal opportunities that might be less obvious, but just as significant. Check the time for each operation on each part, looking for bottlenecks in any process. Let your experience tell you which times seem unreasonable and which methods seem inefficient. Then consider the approaches offered in this guide.

Pneumatics Automation as a Cost Reduction Tool

Your company is only as strong as its ability to compete. By learning how to reduce the cost of producing your products, you will become an even more valuable member of your manufacturing team and your company will become an efficient and competitive producer.

Once you understand how to use pneumatics automation, your dependence on the large capital expenditures needed to increase efficiency should decline. Because of the low cost of pneumatics automation, even job runs in the hundreds of pieces can be automated with excellent results, not only financially, but in efficiencies gained.

Now for some specifics. *Pneumatics* refers to an interrelated group of automation building blocks that use compressed air as a power source. You can use these components to assemble your own economical, cost-saving devices. Choose from devices including:

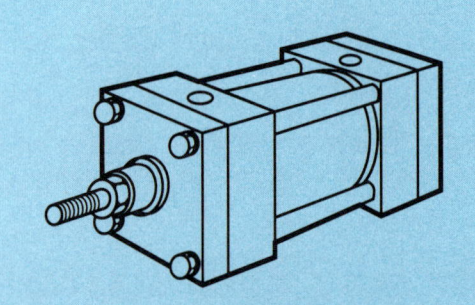

- **Motion and work-generating products** such as cylinders and rotary actuators, automatic drills, reciprocating and rotary work feeders, power presses, grippers, vises, air-powered collets and air motors.

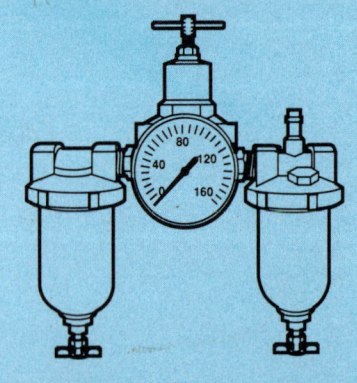

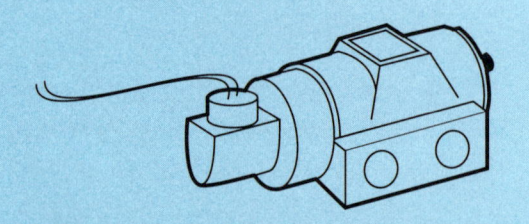

- **Control products** such as directional valves, sensors, logic valves and programmers, all of which are capable of responding to electronic or other forms of input. There also are motion controls such as flow control valves, quick exhaust valves and shock absorbers.

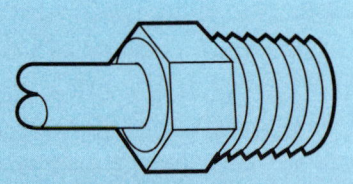

- **Air line treatment devices** like filters, regulators, lubricators and dryers that assure a constant flow of high quality compressed air.

- **Connection products** such as fittings and tubing that link power and control elements.

- **Accessory products** such as pressure boosters, air-over-oil devices and vacuum generators.

Pneumatics Automation as a Cost Reduction Tool

Automation with pneumatics can save you money. We have identified six applications in this guide, but there are many more. Just to get you started, we'll show you how pneumatics will enable your manufactured parts to be . . .

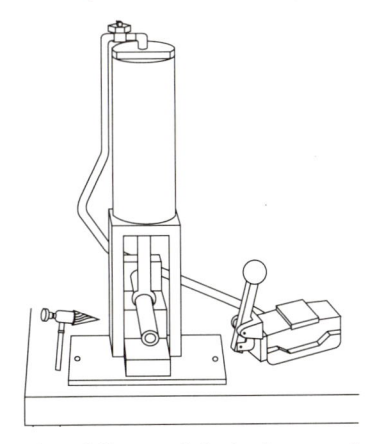

- held firmly while work is being performed.

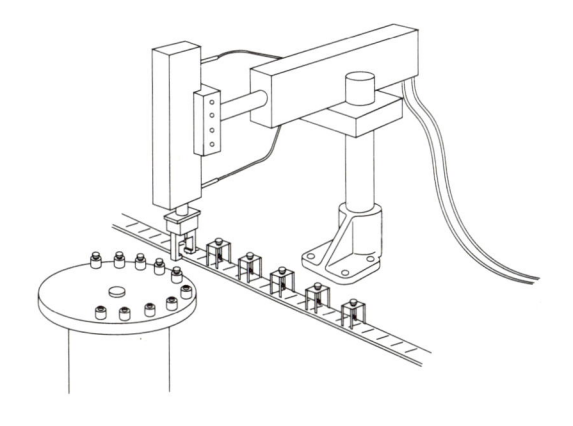

- assembled at maximum speed.

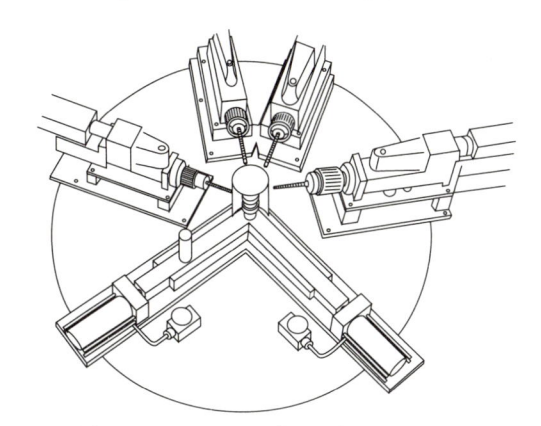

- drilled from many angles at once.

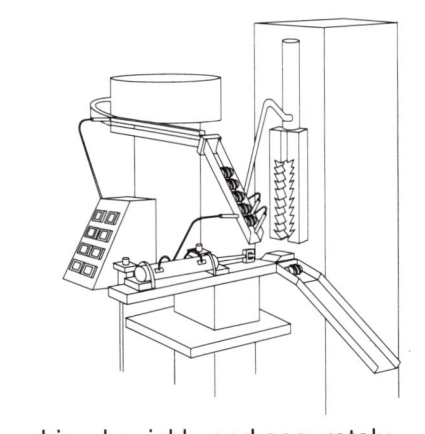

- machined quickly and accurately.

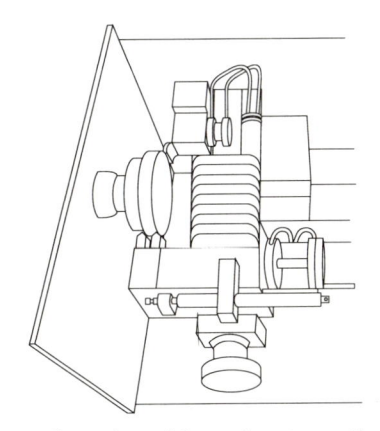

- chamfered and faced automatically.

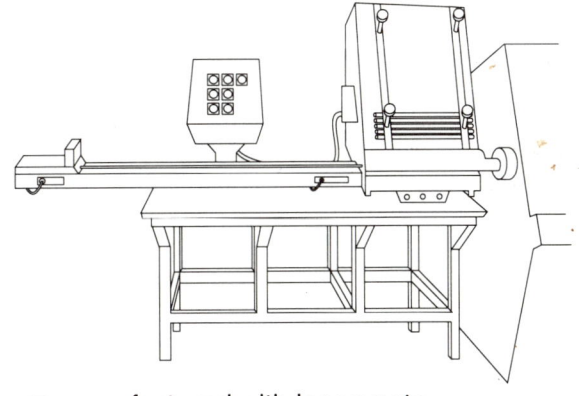

- manufactured with less waste.

Pneumatics Automation at Work – Clamping

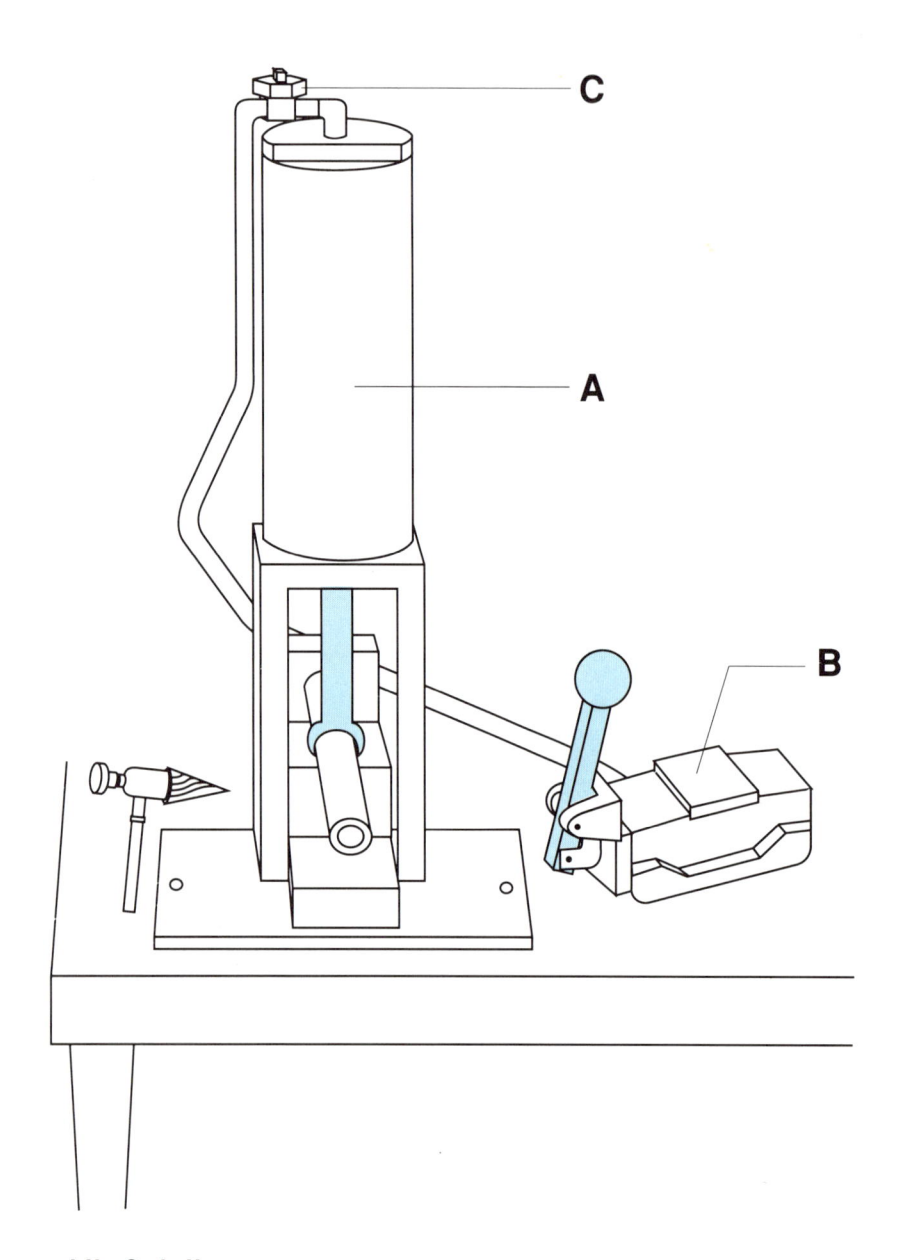

The Problem and Its Solution

A furniture manufacturer produced a specialized tubular steel utility cart made to carry an electronic medical diagnostic instrument. The product was produced infrequently and at such low quantities that dedicated machinery to cut and deburr the support tubing was not cost-effective. Consequently, each section of tube was cut and deburred manually. This operation required placing leather belting around the tube and inserting it into a combination vise for measuring, cutting and deburring. The deburring was done with a hand-held ratcheting type reamer.

To increase production speed, a vertically-mounted pneumatic cylinder was used to hold the work in a predetermined position. This eliminated the need to manually clamp the tubular steel in a vise and mark it every time a cut was made.

Clamping

The Design and Construction Process

1) First, the approximate amount of twisting force generated by the action of the hand reamer was determined, in this case by attaching a common torque wrench to a piece of tubing which was being reamed.

2) Next, the coefficient of friction was determined for leather against steel. This is the ratio between the normal force (or clamping force) and the amount of force it takes to begin sliding one of these surfaces against the other. This was determined by consulting a machinery text.

3) The torque, which represents the turning force, was divided by the coefficient of friction to determine the necessary clamping force. An appropriate pneumatic cylinder was then selected from the manufacturer's sizing tables. The cylinder was a single-acting spring return configuration. It was of the non-rotating type to ensure that its center rod would be in the same position each time its end tooling contacted the tubing.

4) A yoke clamp was mounted to the cylinder's center rod which was shaped to conform to the surface of the tubular steel. A corresponding channel was cut into an opposing clamp block which held the tube and formed the base of the frame. A stop block and reference line were also attached to this base to eliminate the need to individually measure and mark the tubes for cutting. The yoke clamp and clamp block were lined with leather to resist scuffing.

5) Measurements were made and the cylinder frame was welded together, and in turn welded to the channeled clamp block. A stop was welded to the rear of the channeled clamp block, and a mark where the saw cut was to be made was placed near the channeled clamp block's front.

6) The yoke clamp tooling was formed and bolted to the cylinder's piston rod. The cylinder was then mounted to the cylinder frame. A lever-actuated, two-position, three-way control valve was attached to a quick exhaust valve. This valve was mounted to the cylinder's only port. The lever-actuated control valve was then connected to a compressed air source.

Payback: Manual vs. Automated Processes*

Time Savings Per Unit
45 seconds (manual) - 38 seconds (automated)
= 7 seconds

Labor Savings Per Month
.002 hour (7 seconds) x $10 per hour (average hourly rate) x 11,760 units per month = $235

Number of Months to Payback Investment
$200 component cost / $235 monthly labor savings
= .9

Pneumatic Components Used:

A 1 single-acting, spring return, non-rotating pneumatic cylinder (page 28)

B 1 lever-actuated, two position, three-way directional valve (page 47)

C 1 quick exhaust valve (page 51)

2 fittings (page 35)

tubing (page 35)

*NOTE: Supplied figures for all applications in this guide are based upon 21 work days per month with one 7-hour shift operating and an average hourly rate including benefits but not including operating overhead. The component costs listed do not include the tooling or labor required to build the application. For an estimate of total application costs, double the total component cost. Final application costs will vary based upon individual labor costs, skill levels and final application design.

Pneumatics Automation at Work – Facing

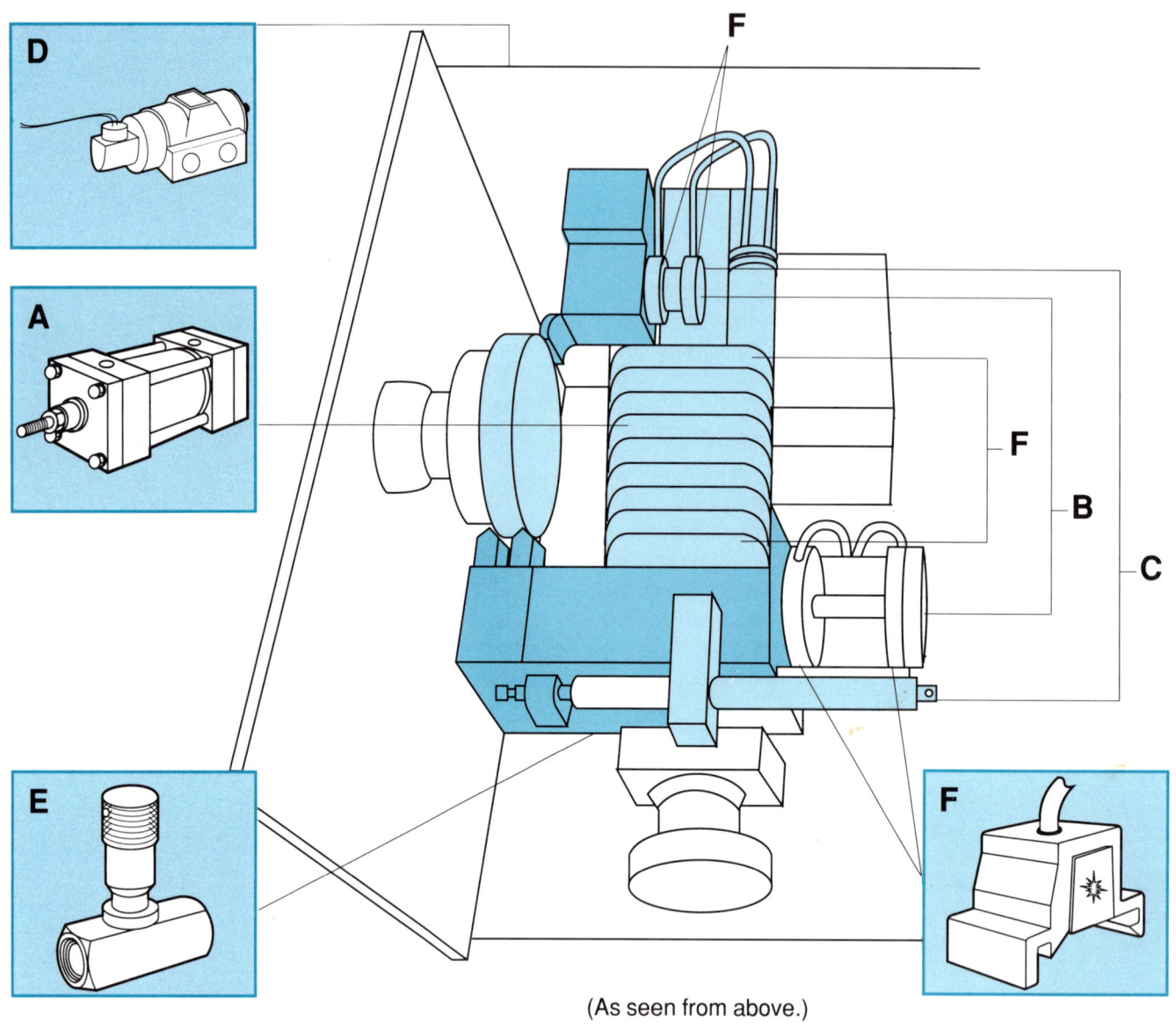

(As seen from above.)

The Problem and Its Solution

A manufacturer made end caps in a multi-operation, single station process. Starting with sections which had been sawed from 5" diameter bar stock, a recessed area was turned into one side of the end cap with a lathe. One of the outside diameters of the end cap was chamfered. The end cap was then reversed and a chamfer was cut along the other diameter. Next, a finishing cut was made along the entire outside diameter from front to back. Tooling was changed, and a finishing cut was made along the face of the end cap.

The lathe operator stood idle during many of these operations. In an effort to increase throughput, a second lathe was installed to execute the chamfering and finish facing cuts. It utilized an automated cross slide which, once started, completed all operations without being attended. This allowed the operator to return to the original lathe to perform the recessing operation which was the most time-consuming phase of the process.

Facing

The Design and Construction Process

1) The cross slide from the original lathe was removed from its saddle, and tooling for a new cross slide was prepared which would accommodate a pneumatic cylinder to provide the forward and backward movement.

2) The appropriate pneumatic cylinder was selected (see clamping application, pp. 6-7 for selection process). A combination air-oil pneumatic cylinder was chosen to ensure smooth action along the cylinder's entire stroke. The pneumatic cylinder was configured to allow four positive stopping positions and featured center rods protruding from both ends.

3) Two sliding tool rests, which moved perpendicularly to the cross slide, were attached to the end of each of the pneumatic cylinder's center rods. Because of space restrictions, ultra-compact pneumatic cylinders were attached to these sliding tool rests to provide the needed motion, which was parallel to the lathe bed.

4) On the forward tool rest, a cutting tool was mounted to provide chamfering and facing of the end cap's outside diameter. On the rear tool rest, a cutting tool was mounted to provide the finishing cut for the face of the end cap.

5) Each pneumatic cylinder used included a position sensor to mark when the cylinder reached its end position, indicating that particular operation had been completed. Using the on/off signals provided by this sensor, a series of solenoid-operated directional valves and relay switches were constructed to start, stop and reverse the direction of each of these pneumatic cylinders as they performed their tool movement functions. The series of operations was started by closing a manual switch. Rates for pneumatic cylinder movement were metered by a series of flow controls between the directional valves and the cylinders.

6) Hose, fittings, wiring connections and control mounting were added. A quick release chuck was attached to the lathe's spindle to facilitate the loading and unloading of the end caps. The control system was connected to the lathe's motor to start and stop the spindle rotation at the appropriate time.

Payback: Manual vs. Automated Processes*

Time Savings Per Unit
3 minutes (manual) - 2 minutes, 15 seconds (automated)
= 45 seconds

Labor Savings Per Month
.0125 hour (45 seconds) x $10 per hour (average hourly rate) x 2,940 units per month = $368

Number of Months to Payback Investment
$4500 component cost / $368 monthly labor savings
= 12.2

*NOTE: Supplied figures for all applications in this guide are based upon 21 work days per month with one 7-hour shift operating and an average hourly rate including benefits but not including operating overhead. The component costs listed do not include the tooling or labor required to build the application. For an estimate of total application costs, double the total component cost. Final application costs will vary based upon individual labor costs, skill levels and final application design.

Pneumatic Components Used:

A 1 single-acting, spring return, air-oil pneumatic cylinder
(page 28)

B 2 ultra-compact pneumatic cylinders
(page 28)

C 2 oil-filled, dampening shock absorbers
(page 43)

D 4 solenoid actuated, two-position, two-way directional valves
(page 47)

E 8 flow control valves
(page 48)

F 8 position sensors
(page 42)

16 fittings
(page 35)

tubing
(page 35)

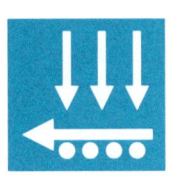

Pneumatics Automation at Work – Assembly

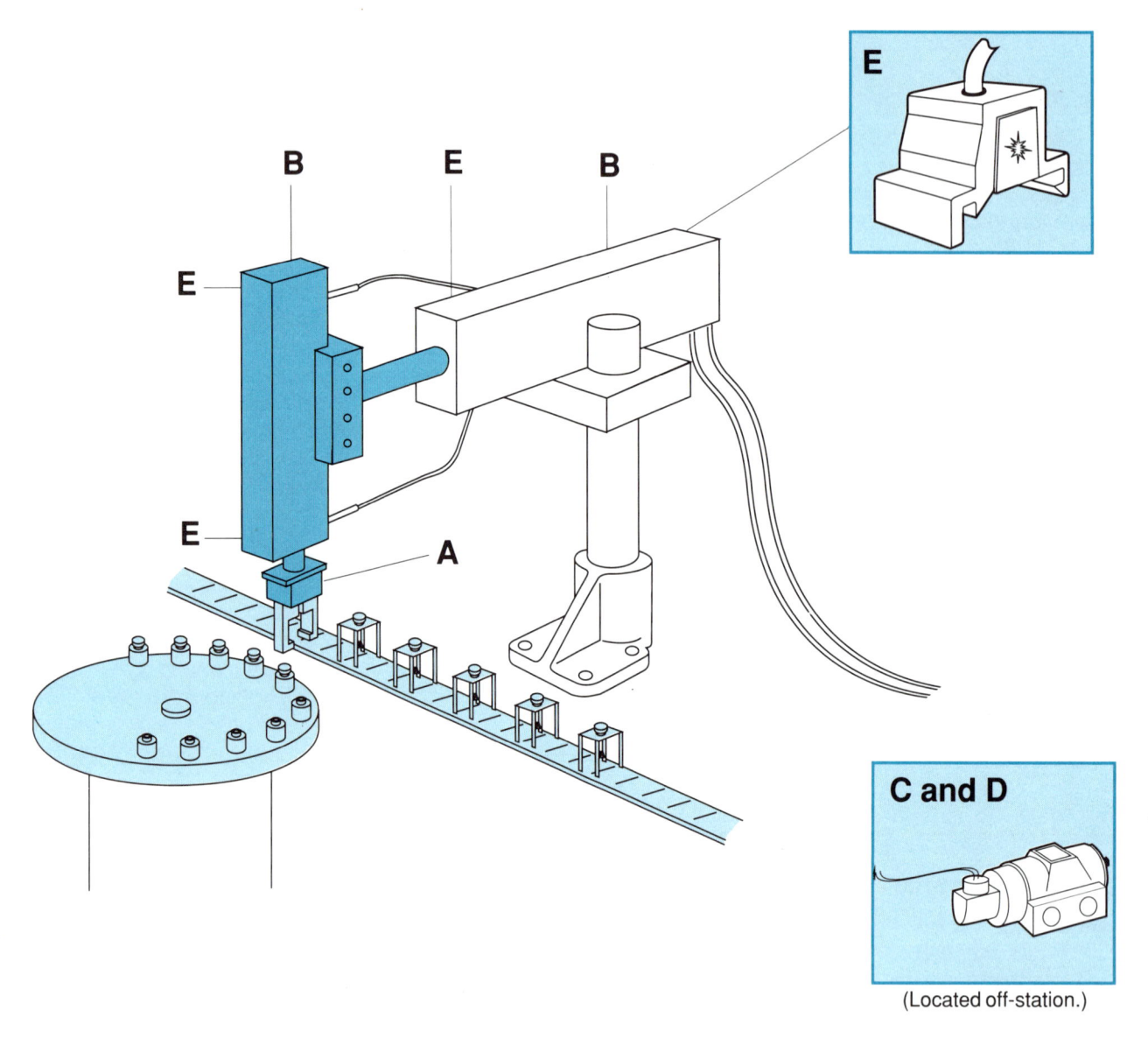

(Located off-station.)

The Problem and its Solution

A cosmetic manufacturer had been manually inserting a brush assembly into nail polish bottles. The process required a worker to pick up each brush individually and insert it into the open nail polish bottle. The bottles were transported to and from the assembly area via a conveyor system. The bottle caps were seated and twisted onto the bottles at a later assembly station. In addition to being labor intensive, the process required a high degree of individual accuracy in order to fit the brush tips into the comparatively small hole in the nail polish bottle. This greatly limited production line speeds.

It was decided to automate this brush assembly process. A gripper was used to grasp the brush. Two pneumatic cylinders then raised the gripper holding the brush, moved it over the open nail polish bottle, lowered the brush into the bottle and released it. The caps were again put on and tightened at a later station. The pneumatic cylinders employed adjustable internal stops so that different sizes of bottles and brushes could be used on the same assembly line. An indexing table, instead of a conveyor system, was used to transport the bottle to the next station.

Assembly

The Design and Construction Process

1) A study was made to decide what motions would be required. The number of cycles the machinery would need to withstand during its lifetime was determined along with the minimum cycle time for this particular operation. Other factors such as maximum allowable "down time" due to component wear and reliability were also examined. In this case, it was decided to use components rated for maximum life, since machine "down time" needed to be kept at an absolute minimum.

2) Components were selected. A low-force gripper, with less then 10 pounds of force, was selected to pick up the brush. This was to ensure that the tubes forming the brush's body would not be crushed during the pick up operation. Special tooling was constructed for the gripper's fingers in order to grasp the brush's body. The weight of the gripper and tooling were used as a means to select the pneumatic cylinder which would provide the vertical or x-axis movement. The weight of this cylinder and the gripper were then used to select the pneumatic cylinders providing the horizontal or y-axis movement of the process. It was decided to use internally supported, non-rotating pneumatic cylinders in order to provide more lateral strength and accuracy. These cylinders included internal stroke adjustments, allowing their use with varying height bottles.

3) Directional valves to control the pneumatic cylinder and the gripper jaws were selected. It was decided to employ three-position solenoid actuated directional valves to control the pneumatic cylinders in order to positively shut off air flow when the assembly station was to be serviced.

4) Measurements were made for locating the entire assembly at the correct point on the assembly line. The mounting kit provided by the pneumatic cylinder manufacturer was selected because it allowed the components to be adjusted after assembly.

5) Pneumatic and electrical connections were made between the compressed air source, valves, pneumatic cylinders, gripper and an existing programmable controller. The controller's internal program was adapted to sequence the assembly station, receiving inputs from sensors located at the ends of the pneumatic cylinders and in the gripper. These sensors indicated whether a particular operation had taken place. Flow controls, connected to the pneumatic cylinders, were adjusted to attain the proper cycling speeds.

Payback: Manual vs. Automated Processes*

Time Savings Per Unit:
 1.7 seconds (manual) − 1.2 seconds (automated)
 = .5

Labor Savings Per Month:
 .00014 hour (.5 seconds) x $10 per hour (average hourly rate) x 311,294 units per month = $436

Number of Months to Payback Investment:
 $2,700 component cost / $436 monthly labor savings
 = 6.2

*NOTE: Supplied figures for all applications in this guide are based upon 21 work days per month with one 7-hour shift operating and an average hourly rate including benefits but not including operating overhead. The component costs listed do not include the tooling or labor required to build the application. For an estimate of total application costs, double the total component cost. Final application costs will vary based upon individual labor costs, skill levels and final application design.

Pneumatic Components Used:

A 1 single-acting, spring return gripper (page 36)

B 2 double-acting cylinders with internal shock absorbers, proximity sensors and stroke adjustments (page 28)

C 1 two-position, three-way solenoid-actuated directional valve (page 47)

D 2 three-position, five-way solenoid-actuated directional valves (page 47)

E 4 position sensors (page 42)

22 fittings (page 35)

tubing (page 35)

Pneumatics Automation at Work – Drilling

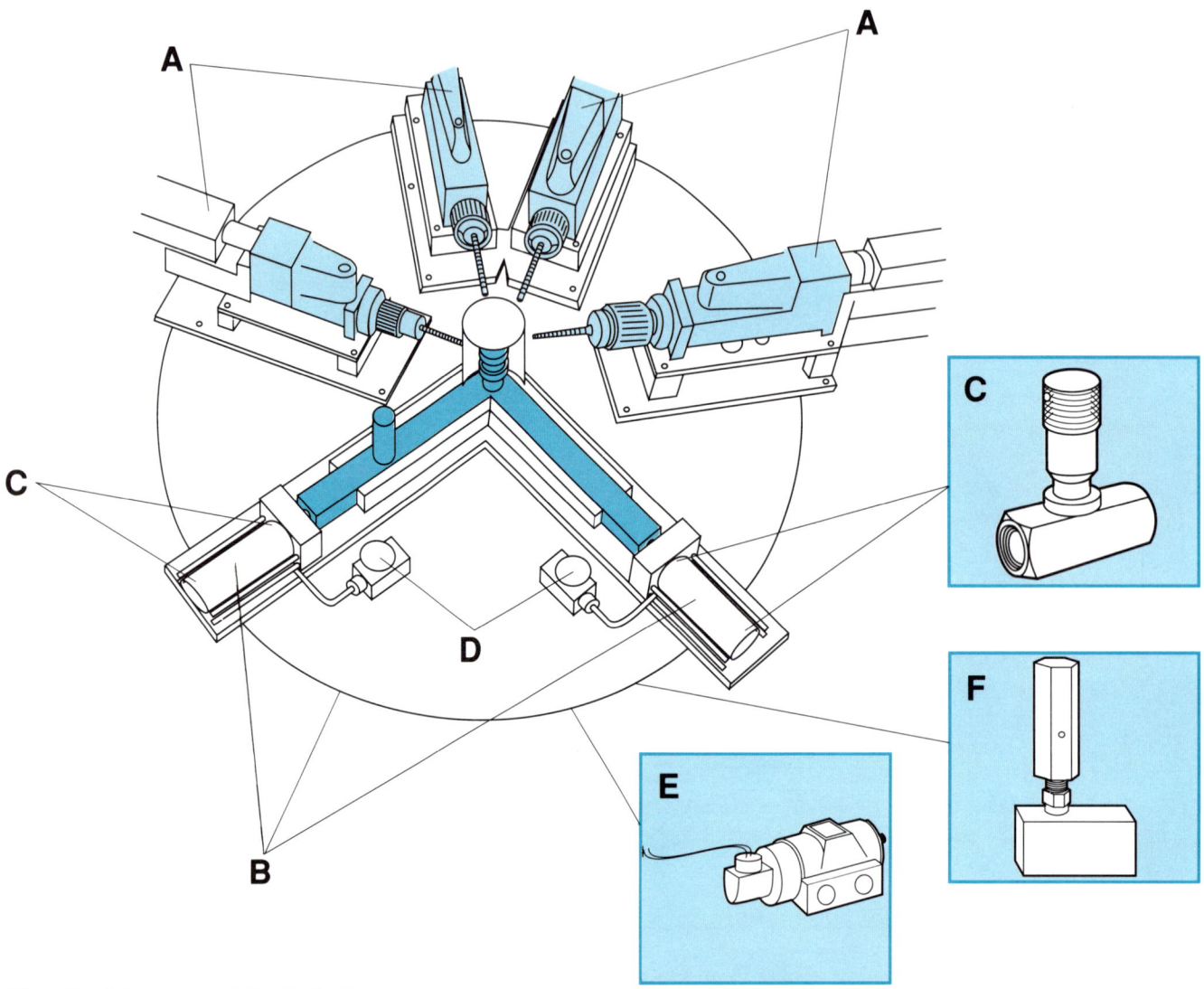

The Problem and Its Solution

A valve manufacturer was drilling four holes in a cylindrical-shaped aluminum sleeve, and then manually deburring those holes with a wire wheel. The holes were drilled one at a time using a standard drill press and a fixture. The holes were drilled at approximately every 90 degrees around the perimeter of the sleeve, with two holes being drilled in one plane, while the other two holes were drilled in another plane.

The manufacturer decided to design a dedicated station for this operation which utilized four self-propelled air drills placed horizontally on a bench in a circular pattern, with the drill's chucks pointing toward the center. The process started when an operator depressed a manual valve. One of two powered sliding pegs carried a sleeve toward the center. Once in the center, a pneumatic cylinder then pushed the sleeve upward, wedging it against an overhead plate. Two of the drills would then simultaneously enter in a different plane, boring through the wall of the sleeve. These drills would then retract, and the second set of drills would enter, performing the same operation from different angles. While this drilling operation was proceeding, the operator would load the second peg with a sleeve. As soon as the drilling operation was completed on the first sleeve, its holding peg would retract and the operator would depress a second manual valve, sending the second peg and sleeve to the center. The drilling operation would repeat while the operator manually deburred the first peg. The entire cycle was then repeated.

Drilling

The Design and Construction Process

1) A basic layout was devised indicating a horizontal drilling operation with tandem feeding pegs to hold the sleeves. Approximate cycle time requirements and number of operations per year were established.

2) Major components were selected including suitable air drills featuring self-contained feeding mechanisms which allowed them to be automatically extended or retracted. It was determined to use pneumatic-based switching, or logic devices, to provide the sequencing and timing for the various operations instead of electrical relays or programmable controllers. This meant directional valves switched by air. These directional valves, in turn, controlled the drills and valve sleeve peg motion as well as the under bench pneumatic cylinder motion. Pneumatic valves were also used for timing and speed control.

3) The valve sequencing and timing pattern was diagramed. Outputs from pneumatic sensors indicating end of stroke positions in the pneumatic cylinders and air drills were used to switch the position of the pilots in the directional control valves. Impulse valves were added allowing a pilot valve to change positions.

4) Machining of the bench and peg slides was done next, with mountings provided for the air drills and the under bench pneumatic cylinder which locked the sleeve in place. A safety cut-off valve was added. Air lines were connected to all valves and timing was calibrated.

Payback: Manual vs. Automated Processes*

Time Savings Per Unit
114 seconds (manual) - 32 seconds (automated) = 82 seconds

Labor Savings Per Month
.023 hour (82 seconds) x $10 per hour (average hourly rate) x 4,642 units per month = $1,068

Number of Months to Payback Investment
$4,700 component cost / $1,068 monthly labor savings = 4.4

*NOTE: Supplied figures for all applications in this guide are based upon 21 work days per month with one 7-hour shift operating and an average hourly rate including benefits but not including operating overhead. The component costs listed do not include the tooling or labor required to build the application. For an estimate of total application costs, double the total component cost. Final application costs will vary based upon individual labor costs, skill levels and final application design.

Pneumatic Components Used:

A 4 air self-propelled drills, with internal slides and sensors (page 24)

B 3 double-acting pneumatic cylinders (page 28)

C 4 flow controls (page 48)

D 3 two-way, two-position, manually actuated directional valves (page 47)

E 4 four-way, three-position, air pilot actuated directional valves (page 47)

F 2 impulse relay valves (page 51)

tubing (page 35)

65 fittings (page 35)

Pneumatics Automation at Work – Broaching

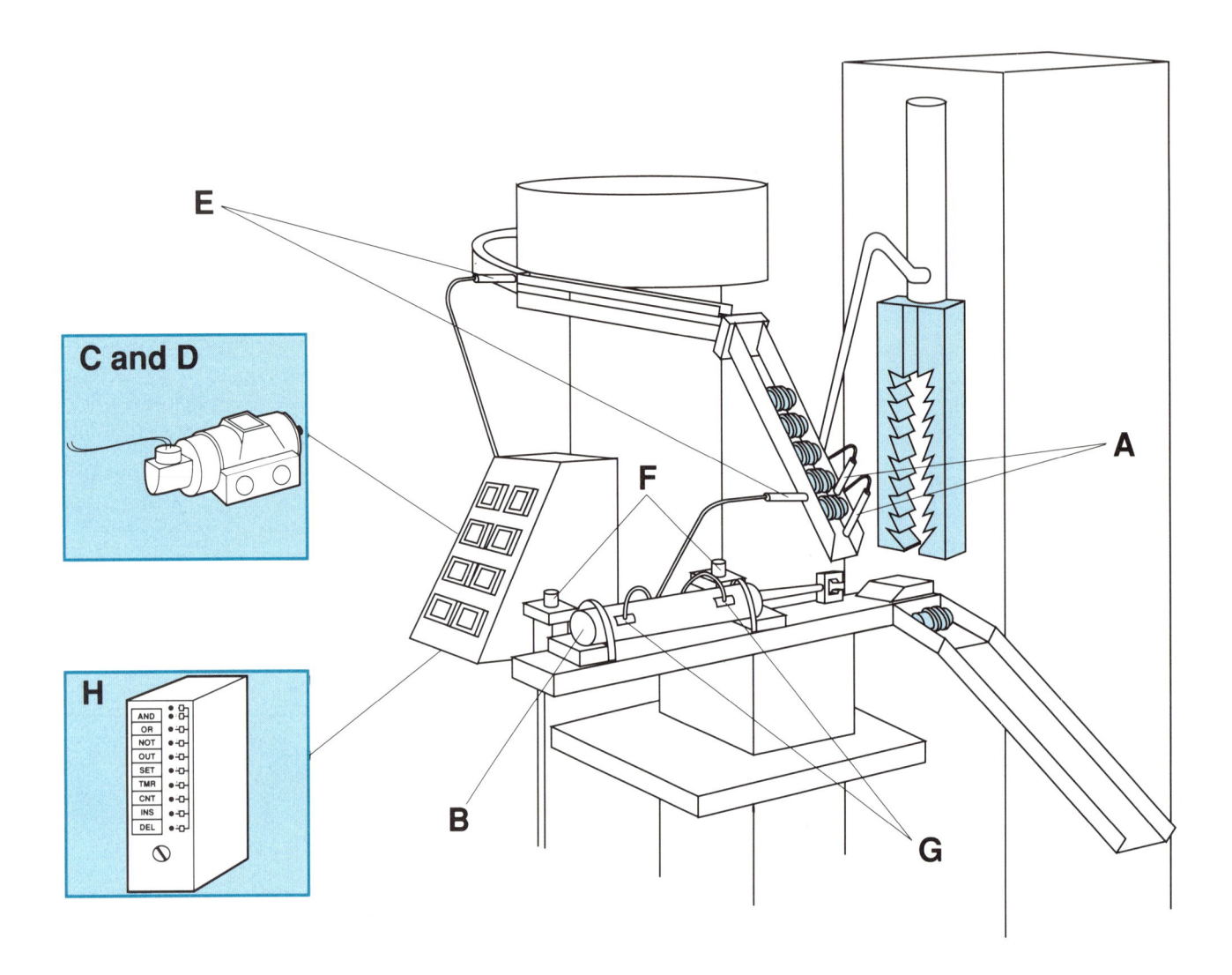

The Problem and Its Solution

A manufacturer made an aluminum fitting as part of a subassembly. It used a horizontal mill to cut four flats into the cylindrical fitting. This process required using a fixture to hold a group of these fittings in line on the bed of a horizontal milling machine. The fittings were moved past a milling cutter in one direction to make flats on one side, and then were turned over for a second milling cut. When the process was completed, the fixture was filled with a new set of fittings to be processed. The machining required an operator to be in attendance at all times.

To increase output from this operation, it was decided to employ a vertical surface broach which would automatically be fed the fittings. The broach used four sets of saw-like teeth to cut all four flats simultaneously. In addition, a vibratory feeder bowl was attached to the broach which allowed an operator to fill the bowl with fittings and then leave the machine in operation to perform another function.

14

Broaching

The Design and Construction Process

1) A standard vertical broach was outfitted with a single-acting hydraulic ram to drive the broaching jaws.

2) A vibratory feeder bowl was mounted to the broach. It included a gravity-fed chute which presented the cylindrical fittings to a mechanical gripper. Proximity sensors were attached at each end of the chute to confirm that fittings were moving through. Two small pneumatic cylinders were placed in line as stops at the end of the chute. This was done in order to present a single fitting to the mechanical gripper. This procedure was accomplished by extending one cylinder into the chute in front of the second to last fitting, halting this fitting, as a second cylinder was retracted to let the last fitting through.

3) The mechanical gripper was attached to a pneumatic cylinder which moved the fitting into the broach. A fixture was placed underneath the lower gripper jaw so that as it was moved by the cylinder, it would hold the fitting tightly against the upper gripper jaw. Two reed-type position sensors were placed at both ends of the pneumatic cylinder to confirm when the gripper was in either the loading or holding (forward or back) position. Flow controls were used to adjust the rate of the pneumatic cylinder's motion.

4) An outlet shoot was attached to catch the broached fittings as they slipped from the loosened gripper jaws during the gripper's retraction stroke. Mechanical trip switches at each end of the broach's stroke signaled its readiness for, or completion of, its cut. A coolant spray, circulation pump and catch basin also were utilized, as were solenoid-actuated pneumatic and hydraulic directional valves to control the action of the pneumatic cylinders and the hydraulic ram.

5) A programmable controller sequenced the entire operation. It used the inputs from the proximity and position sensors and trip switches to determine when to signal the start of the next phase in the operation. Operation ceased when a sensor failed to signal the completion of a cylinder's stroke or found no fittings at either end of the supply chute. The controller was wired into an operator control panel.

Payback: Manual vs. Automated Processes*

Time Savings Per Unit
13 seconds (manual) - .5 seconds (automated) = 12.5 seconds

Labor Savings Per Month
.0035 hour (12.5 seconds) x $10 per hour (average hourly rate) x 40,708 units per month = $1,425

Number of Months to Payback Investment
$20,000 component cost / $1,425 monthly labor savings = 14.0

*NOTE: Supplied figures for all applications in this guide are based upon 21 work days per month with one 7-hour shift operating and an average hourly rate including benefits but not including operating overhead. The component costs listed do not include the tooling or labor required to build the application. For an estimate of total application costs, double the total component cost. Final application costs will vary based upon individual labor costs, skill levels and final application design.

Pneumatic Components Used:

A 2 single-acting, spring return pneumatic cylinders (page 28)

B 1 double-acting, pneumatic cylinder (page 28)

C 1 two-position, three-way, solenoid actuated directional valve (page 47)

D 2 two-position, two-way, solenoid actuated directional valves (page 47)

E 2 proximity switches (page 42)

F 2 flow control valves (page 48)

G 2 position sensors (page 42)

H 1 programmable controller (page 39)

11 fittings (page 35)

tubing (page 35)

Pneumatics Automation at Work – Loading

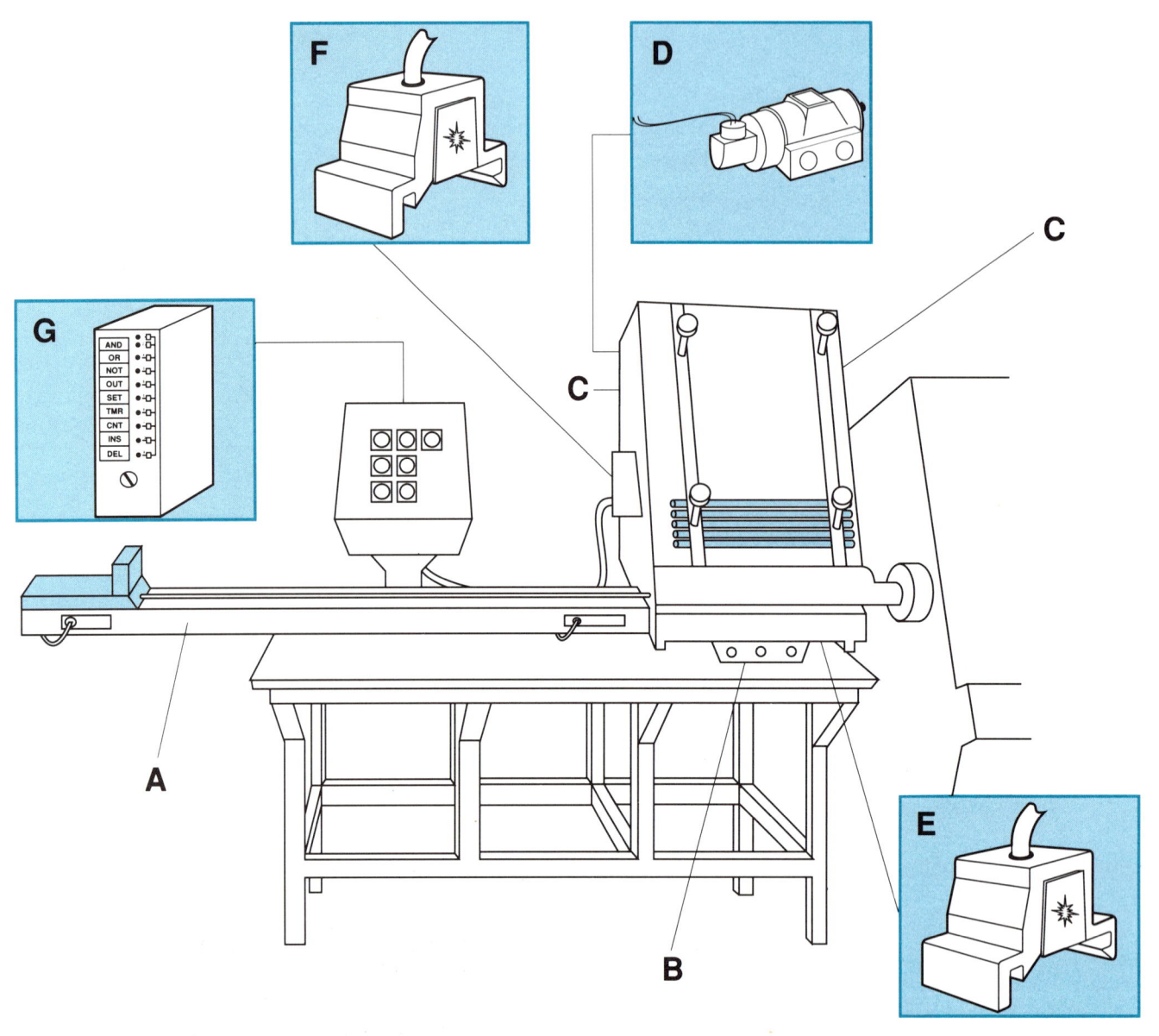

The Problem and Its Solution

A manufacturer wanted to be able to utilize the remnant bar stock left over from larger lathes which had a 15" minimum usable bar stock length. However, the set up and feed time required to utilize these shorter stock lengths was prohibitive since it required an operator to hand load, register and feed the stock in a different machine after each finished part was cut off.

In order to solve this problem, the manufacturer decided to add an automated magazine to one of its smaller, rear spindle-fed lathes. This would allow an operator to fill the magazine with up to 30 pieces of approximately the same size remnant bar stock. The magazine would then present one piece of remnant stock at a time, while a pneumatic cylinder would load and register the remnant into the air collet on the lathe. Sensors in the automated magazine would then signal the lathe to start the machining process, and advance the remnant as needed. The lathe automatically expelled the finished parts.

Loading

The Design and Construction Process

1) An adjustable bar stock loading magazine was designed and built which could accommodate round bar stock from 1/4 to 1-1/4 inches in diameter, and from 7 to 15 inches in length. The design called for the bar stock to be within plus or minus 1 inch of the same length. The magazine was fitted with two devices known as escapements. These contain twin pneumatic cylinders which alternately extend and retract to allow only one object at a time to pass.

2) A vertically adjustable V-block bed was made to accept each piece of remnant stock. The vertical adjustment allowed the V-block to accept varying diameters of stock, while always staying level with the center line of the lathe's spindle. A rodless pneumatic cylinder was attached to the V-block to push the stock into the lathe's spindle.

3) A pressure sensor, attached to the rodless pneumatic cylinder, signaled when the bar stock had reached a stop block on the lathe's turret. The system was designed so that when this stopping action occurred, the lathe's collet closed and the cylinder retracted. In addition, a fiber optic-based sensor installed near the V-block indicated whether a part was present or not.

4) A pneumatic cylinder, made into a slide, was mounted to the bed of the assembly in order to move the entire loading mechanism out of the way should manual access to the lathe be necessary.

5) Three solenoid actuated directional valves were added to control the escapements, the rodless pneumatic cylinder and the slide style pneumatic cylinder. Wired to the valves, the sensors and the lathe, the programmable controller started the remnant loading process upon the receipt of a "chuck open" signal from the lathe. The controller was programmed to accept a signal from the lathe indicating the machining process was complete, whereupon it advanced the existing remnant further into the collet or loaded a new one. The process was then repeated.

Payback: Manual vs. Automated Processes*

Time Savings Per Unit
240 seconds (manual) - 9 seconds (automated)
= 231 seconds

Labor Savings Per Month
.064 hour (231 seconds) x $10 per hour (average hourly rate) x 2,205 units per month = $1,411

Number of Months to Payback Investment
$6000 component cost / $1,411 monthly labor savings = 4.3

*NOTE: Supplied figures for all applications in this guide are based upon 21 work days per month with one 7-hour shift operating and an average hourly rate including benefits but not including operating overhead. The component costs listed do not include the tooling or labor required to build the application. For an estimate of total application costs, double the total component cost. Final application costs will vary based upon individual labor costs, skill levels and final application design.

Pneumatic Components Used:

A 1 rodless pneumatic cylinder (page 28)

B 1 slide style pneumatic cylinder (page 28)

C 2 escapement style cylinders (page 28)

D 3 two-position, four-way, solenoid actuated directional valves (page 47)

E 1 fiber optic sensor (page 42)

F 1 pressure differential switch (page 42)

G 1 programmable controller (page 39)

14 fittings (page 35)

tubing (page 35)

How to Carry Out Your Own Pneumatics Automation Project

In order to construct the best pneumatics automation project, here are the steps you should follow:

■ Determine the environment.

This includes the availability of a constant supply of compressed air operating at an acceptable pressure level (usually around 90 pounds per square inch). Suitable electrical power sources will usually also be required.

Record the operating temperature and moisture level limits to determine if special types of materials, shielding or sealing will be required, and that the components you choose will operate properly under your specific conditions. If the application will receive parts from, or transfer parts to, another assembly station, conveyor line, pallet, parts bin or similar conveyance, establish minimum and maximum rates for parts acceptance or transfer. If your application must interact with another machine (such as switching it on or off), check to see what the control requirements are for that machine in terms of electrical load, cycle rate and availability of connections.

■ Study necessary motion.

Generally define the type of motions necessary to perform the task to be automated by studying the manual processes currently being used. Look for inefficiencies in the manual system, and try not to reproduce them through automated movement. From the point of delivery to the automated station, trace the path of the part, noting where it must be staged, clamped, turned or otherwise manipulated. Try to determine the most direct path for material movement which can be executed by moving the part from one point to another, usually within a single plane.

Routing the part to its ultimate destination usually requires a series of these "point-to-point" moves, since tracing an exact angular path across two planes may require special guides or fixtures. Think of straight-line moves: up, down, over, in or out, and you will most effectively utilize simple pneumatics automation.

■ Examine the part.

Determining the size, weight, composition, shape and surface type of the part to be worked is essential. These factors dictate the type of devices, the mounting approach, the tooling used to grip the part during movement, and the load capacities of the components selected.

If a relatively heavy (over 1 pound) part must be moved horizontally with any degree of accuracy, selection of heavier duty slides versus simple cylinders may be in order. Part surface type and material composition will determine use of gripping device type including angular versus parallel jaw grippers, suction cups or other holding means. Part shape will determine the type of tooling which will be required to make direct contact. This tooling is almost always customized due to the diversity of part sizes and shapes. On heavier parts, some type of deceleration device may be needed when cycle speeds are fast.

■ Determine the part quantities, production speed and accuracy.

If the application is to be used only for a limited run of parts and will then be dismantled, it will endure lighter duty components and less permanent mounting approaches. Production speed will determine compressed air pressure ratings, air flow, parts routing and the need for alternative actuation power sources (such as use of electric devices like stepper and servo motors). If a part must be moved very

How to Carry Out Your Own Pneumatics Automation Project

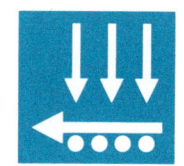

quickly, the accuracy with which it can be moved in a single segment may be lessened.

Accuracy requirements will also determine the need for guide fixtures and jigs, use of non-rotating versus rotating cylinders and sensor selection. Sensors, for instance, are used to determine whether a part has been successfully picked up or moved to a certain point. (If less sensing accuracy is required, a proximity switch may be employed which may sense to within plus or minus .125 of an inch. For increased accuracy, an electronic set point module may be used with a variable output sensor which can achieve sensing accuracies to within .007 of an inch.)

■ Measure the application and select the components.

Once the previous requirements have been established, accurate measurements must be made to determine the starting and stopping points for each motion to be effected. This usually means first drawing a dimensional model on paper. If you are adapting the automation around existing machinery, accurate measurements must be obtained or made for entry and exit points to and from these machines. The distances which are to be traveled by the part must also be accurately established and recorded on your dimensional model or layout. Using these dimensions, previously established tolerances, life cycle requirements and cycle time, the proper components must be selected.

It is important to size the components correctly. Buying a heavier-than-needed component may result in less accuracy, slower movement and longer payback periods. Buying a lighter duty component can result in premature wear or improper functioning of the system. Selection of controlling devices, typically programmable controllers, should be made after evaluating the number of switching and sensing operations to be performed and their sequence. Ease of set-up, use and future expansion capacity should also be considered when selecting a programmable controller.

■ Make the fixtures, mount and assemble.

Most simple automation component systems provide a variety of standard transition plates, mounting brackets, stands, stanchions and other rigging devices. As mentioned, special tooling is usually required to pick up and hold the part to be worked. However, most tooling in these situations can be constructed from flat stock, roll stock or other standard material shapes. Electrical connections and control mounting locations will have to be determined and executed, along with safety cut-off switches, guards and other safety barriers.

■ Debug and calibrate.

Most automation applications will require some degree of debugging and calibration. These operations typically take the form of fine tuning the programmable controller sequencing and timing, or adjusting air flow rates and physical stops for the actuators used in the system. Some repositioning of various items may be required to increase efficiencies. (*A Pneumatics Automation Trouble-Shooting Guide* begins on page 20 of this handbook as an aid to this step.)

■ Safety First

All equipment and components, pneumatic or otherwise, should be installed and operated at all times in accordance with the instructions and recommendations of the manufacturer and supplier. Employees should be properly trained in the appropriate and safe use of equipment. A safe working environment is essential for employee well-being and satisfaction, and makes good business sense.

A Pneumatics Automation Trouble-Shooting Guide

Problem: A cylinder rod is moving erratically during stroking.

Solution: Irregular rod motion could be caused by . . .
- Air pressure input that is too low for the load being moved.
- Too small of a cylinder bore size for the load being moved.
- Side loading on a cylinder rod caused by misalignment of the rod and the load.
- Using flow control valves to meter the incoming air rather than the exhausting air.
- No lubrication.

Problem: Machinery is noisy, particularly at the end of cycles.

Solution: Install a cushioning device. Springs, rubber bumpers, cylinder cushions, deceleration valves, dashpots, feed controls, hydraulic checks, control circuits, industrial shock absorbers and linear decelerators are all devices which offer various forms of cushioning or stroke control.

Problem: Noise is coming from the vacuum pump system which drives the suction cups.

Solution: Vacuum ejector-style pumps of the single chamber type often emit high sound levels. Special mufflers are available which can reduce this noise level. The vacuum pump, if a mechanical type, should be lubricated regularly.

Problem: A horizontally mounted cylinder has worn out prematurely.

Solution: The cylinder may be carrying too much weight. Consider replacing it with a linear slide whose guide shafts and bearings are designed to bear heavier loads which are being moved horizontally.

Problem: Automation seems to "misfire" changing sequence or timing at irregular intervals.

Solution: Electrical noise may be interferring with sensor operation. Use a different power supply for the programmable controller and sensors than is being used to drive heavier duty components (motors, pumps, electrodes, etc.). If this is not possible, investigate resistor-capacitor networks or clamping diodes to condition the current against spikes or electrical noise.

Problem: A cylinder piston rod bends or buckles.

Solution: Probably the piston rod diameter was too small for the amount of thrust and stroke length to which it was subjected. The cylinder sizing calculations should be rechecked and the cylinder should be replaced with one containing the proper size piston rod.

A Pneumatics Automation Trouble-Shooting Guide

Problem: Part of an automation application using a programmable logic control does not work or works improperly.

Solution: Check the following:

- Wires not connected to the proper terminals on the PLC, actuator sensor, or solenoid valve.
- A proximity switch or sensor is improperly placed and is not sensing actuator position correctly.
- The polarity of the power supply is incorrect.
- The program in the PLC is too large and is not executing in time to pick up the sensor and switch inputs from an actuator.
- Wires are broken or a terminal is loose.
- Foreign matter inside an actuator is preventing it from completing its full stroke or rotation.
- The PLC program logic is incorrect.
- The PLC is receiving an electrical input or switching an electrical output which is mismated in terms of voltage, current, frequency or other electrical parameter.

Problem: An electrically powered, solenoid operated valve "burns out."

Solution: Solenoid electromagnet coil burnout could be caused by a voltage mismatch, with either too high or too low a voltage being supplied, or a frequency mismatch. Voltage ranges should typically fall within the +/- 10 percent range. On double solenoid valves which are yoked to the same valve spool, energizing both solenoids at the same time can also cause coil burnout. Cycling a solenoid valve too quickly can result in coil failure. Temperature extremes can also promote coil failure, with high temperatures causing a breakdown in coil wire insulation and cold temperatures causing valve parts to distort and drag. Dirt, oil, moisture or other foreign matter invading the valve may cause it to stick or score, again causing coil burnout. In all situations, correction of the abnormal state should be made and the valve should be repaired or replaced.

Problem: Air appears to be leaking form somewhere in the system.

Solution: Leakage can occur from worn or incorrectly installed actuator and valve seals; worn or incorrectly installed fittings, connectors or hose; or corroded or damaged air storage tanks and supply lines. External leaks can be isolated by listening for the sounds of air escaping, by applying soapy water or similar high surface tension substances to suspected leak areas, by unhooking suspected leaky valves or actuators and operating them independently or by measuring the amount of time portions of the systems take to depressurize.

Pneumatic Components

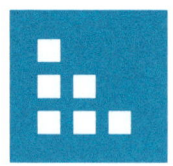

The products pictured are representative of many varieties offered by a number of manufacturers. Check the section titled *Sources for Pneumatic Components* beginning on page 56 for alternative choices.

Please note: Heavy weight arrows indicate mechanical movement while lighter weight arrows indicate air movement.

Air Collets/Chucks

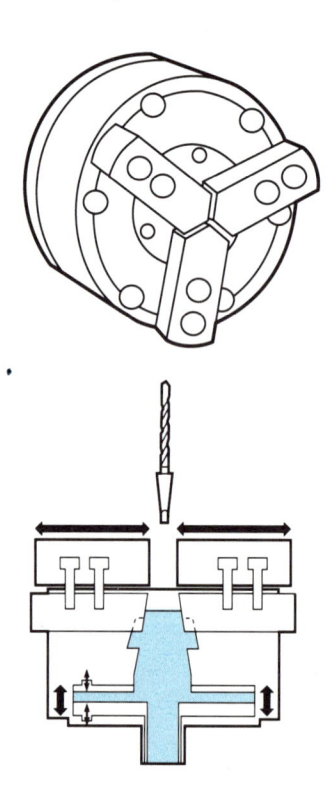

Use:
Air collets, chucks and similar fixtures are used to evenly grip stock during machining, drilling, positioning or assembly with a resultant increase in loading/unloading speed and accuracy.

Operation:
A variety of air chucks and collets exist, but most employ a pneumatic cylinder to drive either a wedge, a metal disk-like diaphragm or a lever. These mechanisms either directly or indirectly contact the jaws of the chuck or collet, forcing them together for gripping the outside diameter of a part, or forcing them apart for inside diameter gripping.

Available Configurations:
- Stationary, for holding work in non-rotating situations.
- Rotating, for holding work in rotating situations.
- Single-acting, depending upon springs to force open jaws for work removal.
- Double-acting, utilizing return cylinder stroke for work removal.
- Safety back-up, providing emergency chucking force in order to prevent work from being thrown from the collet/chuck because of loss of air pressure.
- High force, using dual cylinders for increased force.
- Ultra-precision, offering better tolerances.
- Grinding, with sealed face and master jaw area.
- Dual base, for multiple operations.
- Diaphragm chuck, for precision needs.

Sizing/Selection:
Selection of an air collet/chuck will depend upon mounting requirements, bore capacity (to accommodate various size collets), clamping force and rotational speed.

Calibration/Adjustment:
The collet/chuck is usually equipped with pressure regulation controls so that the pressure on the jaws can be adjusted to prevent crushing of the material being held, or to lessen the strain on the mechanism.

Application Notes:
- Air chucks and collets are available with center holes to accommodate coolant or blasts of air to keep parts cool during machining or to assist in part ejection from the chuck or collet's face.
- Some designs employ an accumulator in their bases so that parts will remain gripped even if the chuck or collet needs to be removed and placed on another machine for a different type of operation.
- Loading pins and rings are available to be used to set the jaw size of the chuck for inside or outside diameter gripping for more accurate set-up.
- Pie-style jaws can be employed in order to grip an irregularly-shaped part and move it to a non-rotating operation.

Pneumatic Components

Air Motors

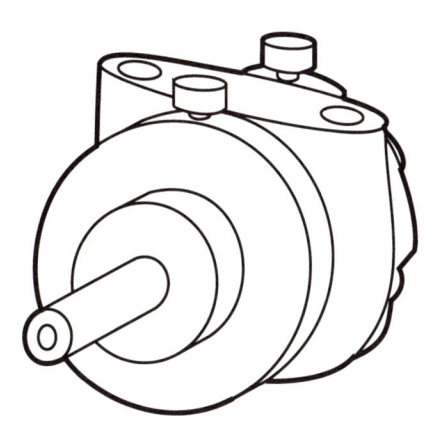

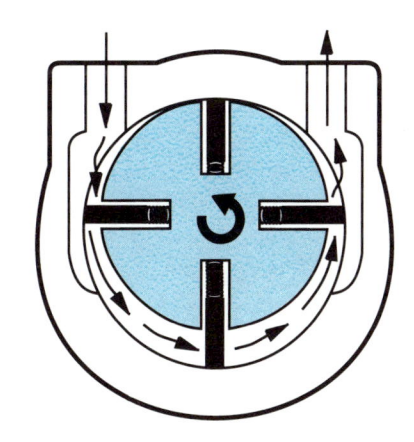

Use:
Used to achieve comparatively high speed, unlimited rotary motion.

Operation:
Two major designs dominate. One is the reciprocating piston type which achieves rotary motion through the timing of the air entering and exiting a piston chamber with a piston rod connected to a crankshaft or cam. The other is the vane type which employs blades longitudinally attached to a rotor which turns when contacted by compressed air. The rotor is typically mounted eccentrically inside the cylinder. Other variations include those which use a diaphragm driving a ratchet and pawl in order to achieve rotary motion, gear motors in which the compressed air presses against the teeth of internal helical gears and turbine motors which utilize internal fan blades set in the compressed air path to achieve a high speed rotary motion. Air motor advantages include high power-to-size ratios, no sparking (important for use around inflammables), self-cooling, an ability to be stalled with no damage, resistance to temperature extremes and being submersible. Depending upon the type, air motors offer a wide range of torque and speed characteristics.

Available Configurations:
- Vane motors – provide low torque and up to 20,000 rpm; often used for air tools, pump drives and light internal combustion engine starting.
- Piston motors – provide medium to high torque and up to 1,500 rpm.
- Turbine motors – provide low to high torque and up to 120,000 rpm; often used for high speed grinders and dental drills.
- Diaphragm motors – provide high torque and up to 100 rpm; often used for valve actuation because of their high breakaway force.
- Gear motors – provide low torque and up to 3,000 rpm.

Sizing/Selection:
The selection and sizing of a gear motor will depend upon the output torque characteristics, operating speed, life cycle, maximum operating horsepower, operating system pressure and mechanical efficiency. Various formulas and tables exist to size an air motor showing the relationship between torque, horsepower, operating rpm, actual operating pressure (pressure differential through the motor) and mechanical efficiency. Consulting the air motor manufacturer for specifications will yield the appropriate selection criteria.

Calibration/Adjustment:
Typically the speed of an air motor is fixed. Gearing is used to reduce motor speed and/or increase torque. Slight speed and torque adjustments can be made by the respective use of flow control valves and pressure regulating valves. Before employing these methods, it is best to check with the motor's manufacturer.

Pneumatic Components

Application Notes:
- Determining the application's need for a reversible versus single direction motor is important since reversible motors are often less efficient than single direction motors.
- Both the output shaft rpm and torque ranges should be considered when selecting motor rpm and gear reduction ratios since extremely high or extremely low motor rpm, when gear reduced, could produce unacceptable amounts of output shaft torque. The speed of a chosen air motor should be evaluated in both loaded and unloaded states; air motors will slow under load.
- Air motors should be operated at the manufacturer's specified pressure, typically with no more than a +/- 10 percent variance.
- Most air motors require air line lubrication and filtration to run properly.

Air Tools

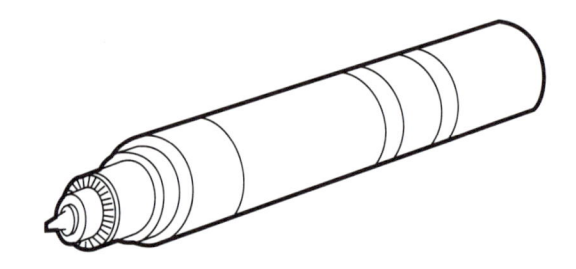

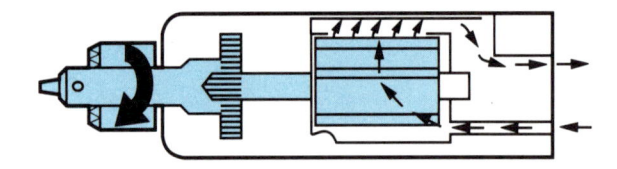

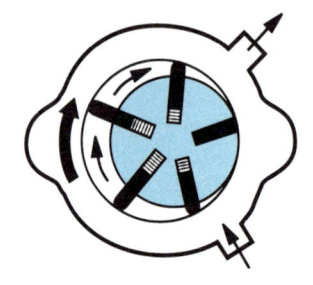

Use:
A miniature form of an air motor, these devices are used to provide rotary force for drilling, screw/nut driving, reaming, tapping and other similar operations.

Operation:
Air drills, screw/nut drivers and similar air tools are typically powered by a vane type of air motor. After entering the tool, a stream of compressed air exerts pressure on the motor's vanes or rotor blades causing the shaft to which they are attached to turn rapidly. The compressed air is usually muffled and then exhausted to the atmosphere immediately upon leaving the tool. Typical reasons for using air tools are higher power-to-weight and size ratio, less maintenance, cooler operating temperatures, less susceptibility to theft, and freedom from the danger of electrical shock.

Available Configurations:
- Reversible tools are able to change rotation direction through gearing changes or air flow reversal.
- Angle or offset tools use additional gears or flexible shafts to be used in areas difficult to reach with a straight shaft tool.
- Wood borers are drills rigged for slower speeds.
- Torque limiting screw/nut drivers allow the operator to preset the amount of torque the device will transmit before the slip-type clutch engages to prevent over-tightening.
- Self-feeding drills are mounted to a drive mechanism, typically either a threaded shaft or a cylinder, to move the device forward as it completes its work.

Sizing/Selection:
Selection of an air tool is dependent upon various application parameters including the rpm, torque, stroke length and speed (for self-feeding models), use of gear reduction or clutch, system operating air pressure, size of chuck or collet to accommodate tooling, and choice of internal or external valving.

Pneumatic Components

Calibration/Adjustment:
Depending upon the application, most of these devices employ some type of gear reduction to slow the tool, or a clutch to stop it, without cutting off the air pressure. Flow controls either in the tool or attached to the compressed air supply line are used to control stroke speed for self-feeding types. Self-feeding types may also include a stroke length adjustment screw.

Application Notes:
- It is advisable to equip certain pneumatic hand tool applications (such as grinding wheels) with the appropriate guards.
- The correct choice and size of tooling should be selected for each pneumatic hand tool to ensure tool longevity and operator safety.
- Include governors on hand tools which will be used with high mass tooling (such as grinding wheels) if there is a danger of tool breakage during operation. Such a tool disintegration could cause the air motor to be damaged by exceeding its rated rpm.
- On self-feeding drills, hydraulic check cylinders can be added to provide a controlled boring rate, even after the bit breaks through the material being drilled.
- Clutches are available which will disengage the pneumatic hand tool when a specified torque level is reached.

Air Vises/Clamps

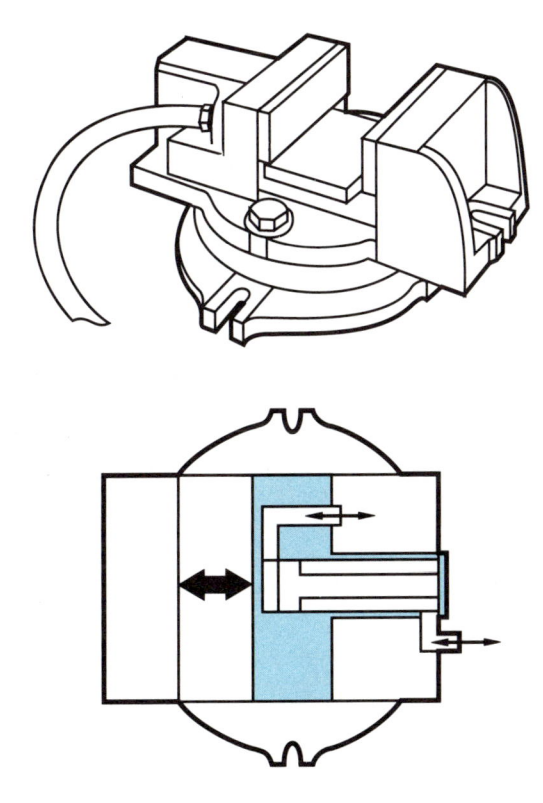

Use:
Provide faster placement and removal of work to be held than with standard manual screw down types of vises and clamps.

Operation:
These devices make use of standard cylinders to effect a clamping force. At standard system operating pressures, clamping forces of 15,000 pounds can be easily achieved without requiring extremely large cylinder bore sizes. In applications using larger cylinder bore sizes with large surface clamping areas, stroke travel may have to be sacrificed to maintain adequate clamping pressure.

Available Configurations:
- Single-acting – provides force in one direction.
- Double-acting – provides force for extension and retraction.
- Air toggle clamps employ a lever attached to a cylinder to increase clamping force.

Pneumatic Components

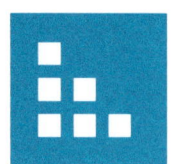

Sizing/Selection:
Proper sizing of an air vise or clamp will depend upon the force it is required to exert, the surface area to be clamped, the speed of its piston motion, the pressure at which it will operate, its life expectancy and its operating environment (i.e., temperature, moisture, maintenance level).

Calibration/Adjustment:
Some air vises or clamps offer adjustments either through a threaded shaft and nut arrangement or through adjustable stop collars. As with most cylinder-based devices, air vises or clamps often come equipped with flow controls that allow metering of the flow of compressed air through the cylinder to change speed. Pressure regulators can control clamping force. Many also contain adjustable end stops and spacers which can be varied to control stroke length.

Application Notes:
- Where higher clamping pressure is needed, boosters or intensifiers may be used (see below). Caution: The cylinder on the vice or clamp must be able to withstand the increased pressure.

- Vise or clamp alignment may be critical during some types of operations. Shims or adjustable bases may be employed to compensate for misalignment.

- Some toggle type clamps will not release with a cut-off of air pressure and so must be pressurized to open and pressurized to close.

- Clamps or vises may be ganged for holding during progressive operations.

- Adequate safety measures such as two-hand interlock switches, photo-electric cut-off switches and guards should be employed during automatic operation. Vise or clamp jaws should be opened no further than necessary for manual part placement to reduce the chance for hand or finger injury.

Boosters/Intensifiers

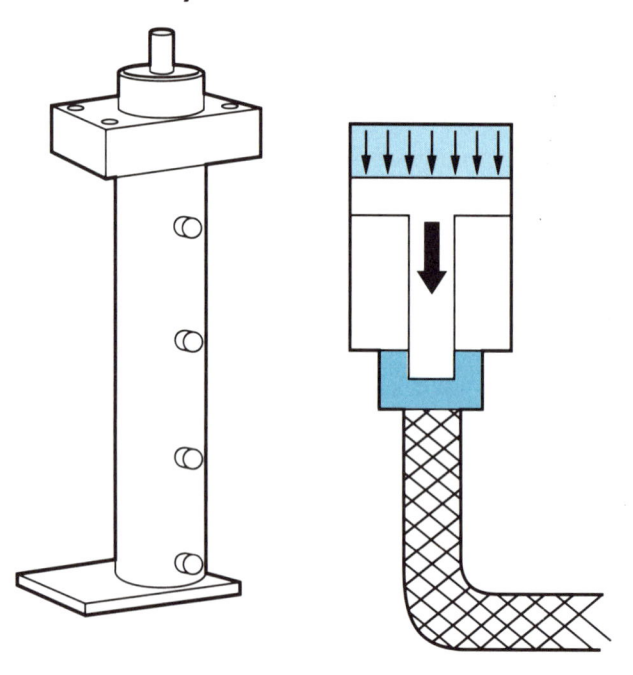

Use:
Used to supply a high pressure output force from a low pressure input force, typically for operations like clamping, punching and die cavity closing.

Operation:
A booster, or intensifier, is essentially a larger piston-type actuator driving a smaller diameter ram or piston rod. This arrangement produces a higher pressure at the end of the smaller piston rod or ram. The output pressure of a booster or intensifier is proportional to the area of the driving piston divided by the area of the driven piston rod or ram. This ratio can be up to 100:1. This means an input pressure of 100 psi acting on a piston with an area of 50 square inches will generate 5000 psi if the smaller piston or ram has an area of 1 square inch. (This disregards frictional losses.) Boosters or intensifiers are used when a relatively small amount of fluid is needed to deliver a higher pressure. In principle, any combination of fluids

Pneumatic Components

can be used on either side of the booster or intensifier. However, typically the booster or intensifier is driven by compressed air and the ram or smaller piston drives hydraulic fluid or contacts the work directly.

Available Configurations:
- Single pressure circuits use one piston to intensify the output which either supplies pressure to another actuator or directly contacts the work area.
- Dual pressure circuits use low pressure to drive a smaller piston which contacts the work, then drive a larger piston into an area behind the first piston, supplying pressure to another actuator or directly contacting the work area.
- Air-oil versions typically use oil as the medium on the high pressure side and air on the driving side.
- Air-air versions, which act as air amplifiers, increasing air pressure at isolated points without the need for additional air compression.
- Single-acting versions rely on springs or another device to restore the booster to its original position.
- Double-ended, double-acting in which intensification is achieved during both the extend and retract strokes.

Sizing/Selection:
Selection of a booster, or intensifier, depends upon the physical size limitations of the area in which it will be installed and the output pressure required. Other factors affecting booster sizing include amount of output fluid required, the pressure at which it will be driven, its life expectancy and its operating environment (i.e., temperature, moisture, maintenance level).

Calibration/Adjustment:
Standard pressure regulators control the pressure produced by a booster.

Application Notes:

AIR-OIL VERSIONS
- Booster-driver cylinders can be used where high pressure is needed within a limited physical space since they can produce the same pressure output as much larger conventional cylinders.
- Boosters can maintain their pressure indefinitely, as long as the booster circuit is not depressurized.
- In general, the use of booster cylinders is limited to simple circuits requiring small to medium volumes of high pressure oil, or short ram strokes.
- Double-acting, or reciprocating-style, booster cylinders can produce larger volumes of high pressure oil than a standard booster cylinder since they provide pressurization on both forward and backward sections of their stroke.
- Booster circuits often contain pressure sensors which cause the booster to operate when a certain pressure level is reached.

AIR-AIR VERSIONS
- Often used in large air distribution systems where points of use are located at great distances from the source of compressed air thus causing pressure drops. These systems often exhibit ample air volume for converting to a higher pressure.
- The force balance principle of construction enables these units to stall at their maximum pressure when there is no flow demand yet automatically cycle at whatever rate is required to satisfy demand at the higher pressure.
- Output flow capacity is inversely proportional to pressure capacity.
- Higher pressure units are commonly used to supply air for production leak testing to reduce the need to purchase nitrogen gas.

Pneumatic Components

Cylinders

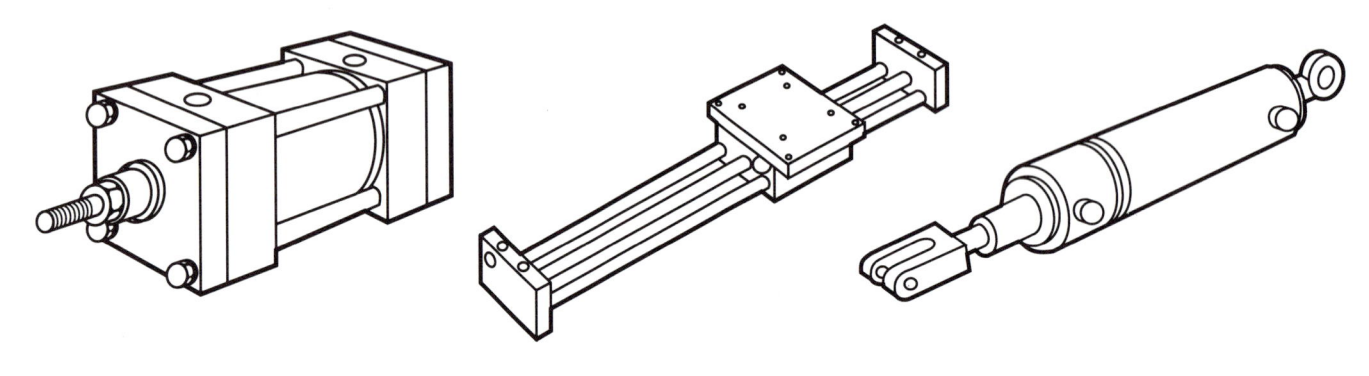

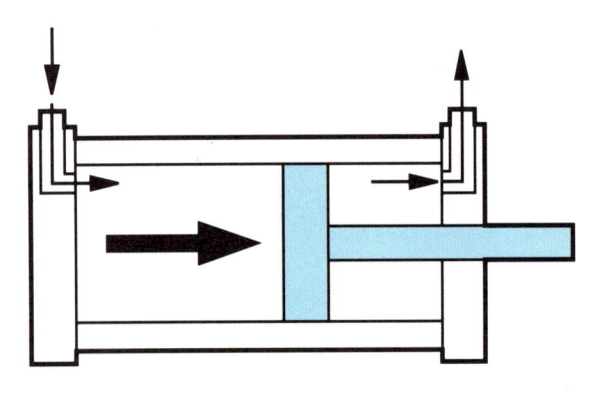

Use:
To provide straightline, "in/out" actuating movement for a wide variety of automation applications in order to move and hold loads or work.

Operation:
Compressed air enters a rigid casing containing a piston surrounded by a seal. The air drives the piston which is connected to a rod extending through the cylinder's end cap. The rod is returned to its original position either by springs or by the introduction of compressed air at the opposite end of the cylinder.

Available Configurations:
- Single-acting – provides force in one direction.
- Double-acting – provides force for extension and retraction.
- Single rod, with a rod extending through one end of the cylinder.
- Double-rod, with a rod extending completely through the cylinder.
- Tandem, with two or more interconnected pistons in the same cylinder providing increased force.
- Duplex, with two independently operating pistons in the same cylinder assembly for multi-position stopping capability.
- Non-rotating – uses a variety of methods to prevent rod rotation during retraction or extension.
- Rodless, which use an internal piston attached to a carrier on the side of the cylinder. These devices are used where space limitations prevent extending a rod beyond the cylinder casing.
- Ultra-compact – used when space limitations or overall length are major design factors.
- Slides – reinforced with guide bars and bearings to increase lateral load-carrying capability.
- Cylinders are frequently offered with built-in or added sensors to indicate end of stroke positions and also with internal cushions.

Sizing/Selection:
Proper sizing of a cylinder will depend upon the force it is required to exert, the speed of its piston motion, the pressure at which it will operate, its life expectancy and its operating environment (i.e., temperature, moisture, maintenance level). Other sizing and selection factors include how much shock the cylinder will be expected to withstand, how much load will be placed on the cylinder's piston rod and bearings and how much drag will be placed upon it from pressure created by the air being expelled from the cylinder.

Pneumatic Components

The application will often dictate one or more of these factors such as available air pressure or load to be moved. Various formulas and tables exist to size a cylinder through determination of other factors. Most cylinder manufacturers have "force and speed" charts which will give this type of information for each of their cylinders.

Calibration/Adjustment:
Cylinders are often equipped with flow controls that allow metering of the flow of compressed air through the cylinder to change speed.

Application Notes:
- There are many different methods for eliminating the need for additional lubrication in cylinders such as special seals, lubricant reservoirs, coatings on the cylinders' walls and internal wicks.

- The output force of a cylinder can be boosted by increasing the inlet pressure to within the manufacturer's recommended limits.

- Excessive force exerted perpendicular to the cylinder piston rod may cause "side loading." The adverse effects of side loading are bearing and seal wear. This can be lessened or eliminated. External guide rails to keep the object being moved in proper alignment may lessen side loading effects. Not using the entire available stroke of a cylinder may also lessen this problem. (This can be accomplished by using internal stop tubes between the piston rod bearing and the piston.) Swiveling rod couplers can also be used in applications where small amounts of misalignment are likely to occur.

- To smooth a cylinder's operation, its piston may be moved by oil under pneumatic pressure, or one section of a duplex style cylinder may be filled with oil which is shuttled in front of or behind the piston as it travels.

Dryers

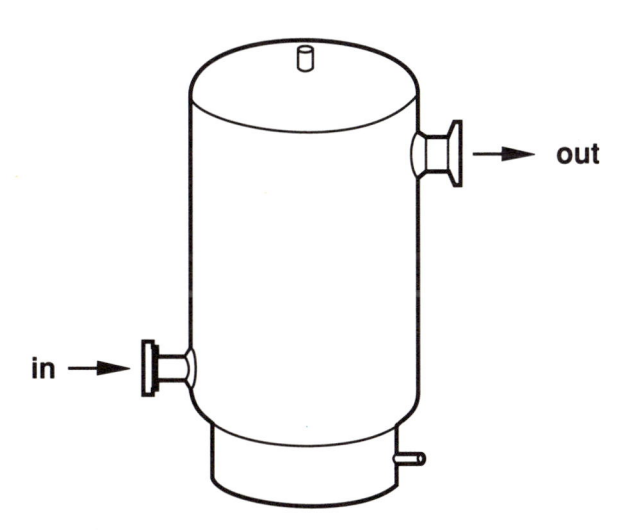

Use:
To extract moisture from a compressed air system by reducing its condensation or dew point.

Operation:
When air is taken into a compressor from the atmosphere, it contains a certain amount of moisture. When this moistened air is used in a pneumatic system, it can condense and deposit water in the air lines and in the components. For certain applications—paint spraying, use of certain air tools, food storage, medical/dental instrumentation, clean room situations and extremely humid or extremely cold ambient environments—drier compressed air may be required. Dryers tend to work either by chilling the air to force the water vapor to condense and drain, or by using a chemical desiccant which absorbs the moisture. Standard filters can remove much of the condensed moisture but they do not lower the dew point. Only if the dew point is low enough can moisture be prevented from condensing in supply lines or components.

Available Configurations:
- Deliquescent chemical dryer systems pass the air through a salt that disolves while it absorbs the water. This will decrease the dew point somewhat below the ambient temperature.

Pneumatic Components

- Refrigeration-based drying systems use a refrigerant in an evaporative system similar to a home refrigerator to cool the air, condensing the water and forcing it out of the system, then warming the dry air back to normal levels for use. These units will decrease a dew point to levels near the freezing point of water.
- Regenerative chemical dryer systems use a chemical desiccant that does not disolve but absorbs moisture. The desiccant must periodically be dried by heat or exposure to dry air. This type of dryer reduces the dew point considerably below the freezing point of water.

Sizing/Selection:
Selection of the type and size of a drying system will depend upon the level of dryness (dew point) desired, the volume of air which needs to be dried, the system operating pressure, the amount of energy consumed in the process, the ambient temperatures, the cost of the equipment and the temperature at which the air is introduced into the drying equipment. Most factory installed compressed air systems should include an aftercooler immediately downstream of the air compressor. This will cause condensation to occur and remove most of the initial moisture. Manufacturers' specification tables for dryers will provide the optimum operating conditions for these types of systems.

Calibration/Adjustment:
Refrigeration-based drying systems, like their residential and commercial system counterparts, contain thermostatic controls which keep their operating temperatures within the proper range. The timing intervals for some types of regenerative-based drying systems can vary depending upon air flow and system pressure parameters. Deliquescent dryer types require daily draining to remove the accumulated water.

Application Notes:
- Caution: Air pressure must be removed from a system before servicing a dryer.
- Dryers should not be located in areas subject to freezing.
- Regular maintenance of a dryer is important to obtain the benefits of dry air.
- Adequate filtration must be installed immediately upstream and downstream of a dryer. Coalescing filters, which remove compressor-generated oil upstream of a dryer, will minimize oil contamination of the chemical media; downstream filtration with standard types will minimize carryover of the chemical media particulates, which can be substantial with high air flows.

Pneumatic Components

Electropneumatic Positioners/Converters

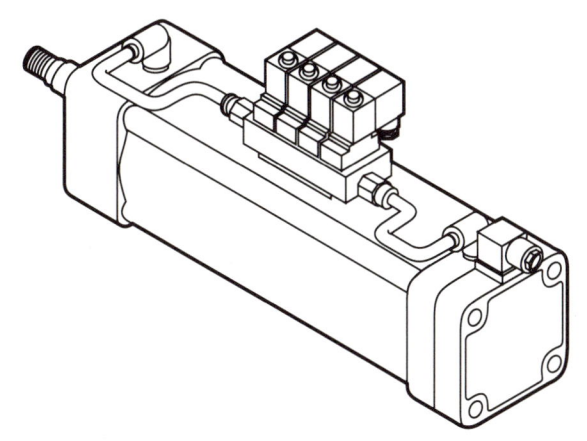

Proportional Control Valves

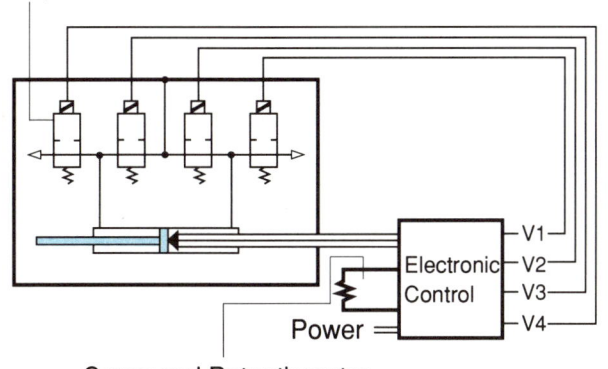

Command Potentiometer

Use:
An electropneumatic (EP) positioner is a pneumatic cylinder controlled by a proportional analog current or voltage signal. It is used for infinite physical positioning action while controlled by some electronic device. An EP converter is a device that controls pressure, delivering an amount of pressure proportionate to an input voltage level.

Operation
An **EP positioner** resembles a standard pneumatic cylinder. It is typically fitted with valves proportionally controlling both the input and exhaust of pressurized air to and from the cylinder. These valves, similar to solenoid driven directional valves, meter the incoming and exhausting air by opening and closing for short periods of time. The number and duration of these "pulses" are controlled by an electronic circuit which is connected to a position sensor contained within the cylinder, and to some type of input control device such as a programmable logic controller or a simple potentiometer. When the input device calls for a specific amount of movement, the microprocessor circuit delivers the appropriate number and duration of valve pulses to move the cylinder the required distance.

An **EP converter** consists of an electronic circuit, 2 two-way solenoid valves, a pressure sensor, a volume chamber and a relay or volume booster valve. After receiving a proportional input, or command signal, from either a programmable logic controller or a potentiometer, the solenoid valves charge or discharge the volume chamber with air. The pressure sensor reports actual delivered pressure via a relative voltage or current output, while a comparator circuit continuously monitors the command signal. A match is maintained between the command signal and the volume chamber pressure by energizing the corresponding intake or exhaust valve.

Available Configurations

-E/P Positioners-
- Single direction positioners, which move from one end of their stroke to the other.
- Available with integral sensors mounted within the positioner's cylinder.

-E/P Analog Converters-
- E/P analog converter which is controlled by a proportional input voltage signal.
- I/P analog converter, which is controlled by a proportional input current signal, rather than a voltage signal.
- Binary E/P digital converter, which depends upon an array of solenoid driven valves operated in various combinations to vary pressure output in discrete steps.

31

Pneumatic Components

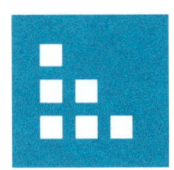

Sizing/Selection:
An E/P positioner is sized according to force requirements, supply pressure, speed requirements, accuracy requirements, environmental considerations, power supply type, input (or command) signal type, feedback requirements and stroke length. E/P converters require a similar set of parameters to be considered when sizing and selecting in addition to delivery pressure and flow requirements.

Calibration/Adjustment:
E/P positioners must be matched to their input, or command signals, and the calibration is usually performed only upon installation. E/P converters are typically factory preset to their rated pressures.

Application Notes:
- E/P positioners may have their comparator circuitry mounted far away from the cylinder positioning element to guard against environmental damage.

- Both E/P positioners and converters are compatible with a wide range of input devices such as PLC's, microcomputers, potentiometers and sequencers.

- Although less accurate than ball screws and stepper motors, E/P positioners are less expensive and may be able to deliver higher positioning forces.

- When using E/P positioners, light loads operating at slow speeds can be positioned more accurately than fast-moving, heavy loads.

- For faster response, the E/P converter should be placed at the controlled device to minimize the fill and exhaust time and pressure drops. (Some applications may require a quick exhaust valve to accelerate the exhaust further.)

- E/P converters should be sized to accommodate any surge in demand. This is accomplished by sizing the supply line accordingly, or by placing a supply volume close to the E/P converter.

Pneumatic Components

Filters, Regulators, Lubricators (FRL's)

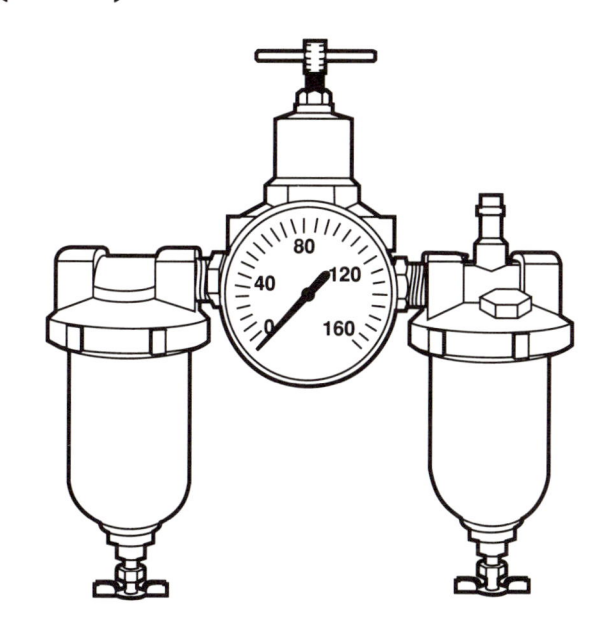

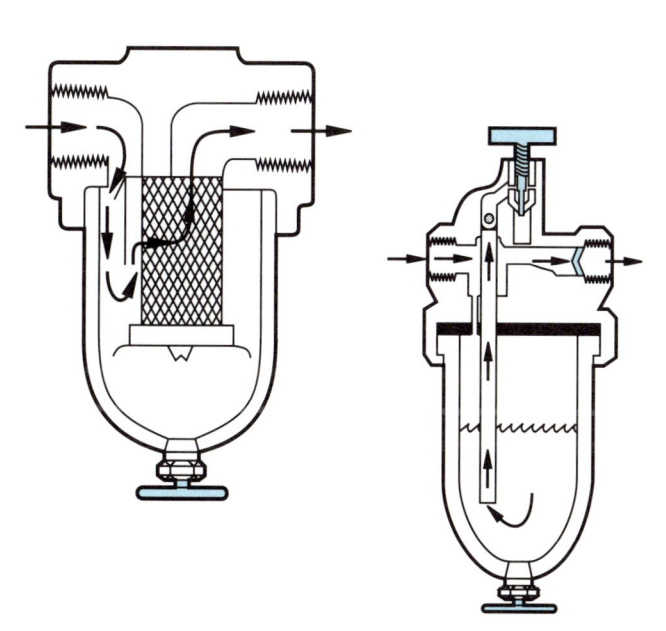

second unit is a pressure regulator (see section titled *Valves, Pressure Regulator*, page 50, for an explanation of this component) which controls the level of outlet pressure, to stay near a certain preset parameter. The third unit is a lubricator which delivers a mist of oil into the outlet air. This lubrication is required by many types of automation components whose seals and internal mechanical parts depend upon the constant presence of lubrication for proper functioning.

Operation:
Compressed air entering the inlet port passes through a deflector inside the **filter**, creating a downward cyclonic flow air pattern. This causes heavier particles and liquids to be propelled against the inside of the bowl and below a baffle which prevents these contaminants from being re-entrained into the air flow. The collected contaminants can be periodically purged via a drain. The air then enters a filter element and is forced through microscopic passageways which trap the smaller contaminants. The filtered air then passes out of the center of the filter. A common type of **lubricator** includes a small venturi located within the main air passageway. The flow of air moving past this venturi creates a low pressure zone at the end of a siphon tube that picks up oil from the lubricator bowl. When this oil enters the low pressure zone, it is regulated by a flexible restrictor and breaks into a fine mist as it joins the main air flow. A check valve is usually included in the pick-up tube to ensure that oil will remain in the line even if air flow stops.

Available Configurations:
-Filters-
- Filters are often distinguished by their elements. An edge-type element is composed of impregnated paper ribbon which catches impurities on the element's outer surface.
- A depth-type filter element is composed of spheres or fibers and the air passes through a tortuous path. This causes filtering action to occur in the porous media.
- Some filter elements are reuseable after cleaning, while others must be discarded after use as stated by the manufacturer.

Use:
Filters, regulators and lubricators are usually connected in one set. This trio is designed to condition and regulate the air prior to its entering a pneumatic component. The first unit is an air line filter which is designed to remove foreign particles and liquids from the supply air. The

Pneumatic Components

-Lubricators-

- Direct injection lubricators depend upon internal pressure differentials and air flow to provide oil for lubrication in mist form. The amount of lubrication is proportional to the amount of air flowing through the system.
- Pulse type lubricators inject a predetermined amount of lubricant directly into a component at a specific time interval without creating a mist.
- A recirculating lubricator also mists the oil, but into much smaller sizes of droplets. This is accomplished by directing the mist from a venturi back into the lubricator bowl where larger particles fall out of the airstream. The fine particles which remain suspended are carried out of the bowl into the main airstream. This allows the lubricator to be placed further away from the device which it lubricates.

Sizing/Selection:

Filter selection is based upon the amount of air flow that passes through and its resulting pressure drop, as measured when the filter media is clean. Filter manufacturers typically provide graphs which indicate these conditions for various component sizes. Similarly, a lubricator is also sized per the air flow rate and allowable internal pressure drop, with selection criteria available from the manufacturer. Both filters and lubricators usually are offered with either clear plastic or metal bowls and with or without sight gauges. Plastic bowls should always be used with a bowl guard because they are subject to rupture.

Calibration/Adjustment:

Since filters are passive devices, little adjustment is necessary. However as the filter element clogs, its resulting pressure drop increases. Lubricators typically employ a needle valve to adjust the flow rate of oil into the air path. Lubricators require periodic replenishment of the proper viscosity and type of oil suited to the particular component being lubricated.

Application Notes:

- Caution: One must be certain that air pressure has been removed and that a system is depressurized on both sides of the filter-regulator-lubricator unit before performing any service or maintenance. Some lubricators may be refilled with oil while the unit is under pressure, but not all types have this capability. Follow the manufacturer's instructions.

- Caution: Clear plastic bowls on filters, lubricators and filter-regulator combinations are sensitive to certain contaminants, oils, cleaning agents and operating environments. The manufacturer's application instructions should be consulted for the correct procedures.

- Never locate filters or lubricators where air flow is reversed during operation.

- Coalescing filters are a special type that remove oil mist (usually compressor oil carryover). The flow direction through a coalescing filter's interior is opposite that of a standard filter. The filter should be installed so that air flow follows the arrow marked on the outside of the unit. The potential for the freezing of the moisture in the filter should be examined before filter placement in a particular location.

- While often mounted with the set screw or adjusting knob pointing downward, a regulator in a filter-regulator-lubricator unit can often be mounted with this adjustment pointing upward.

- Filters should never be selected on the basis of inlet or outlet port size.

Pneumatic Components

Fittings, Connectors, Tubing

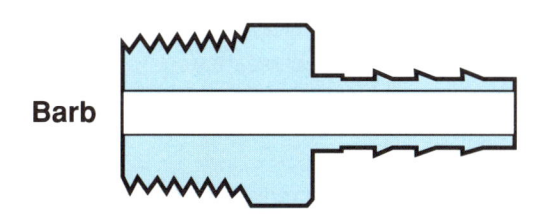

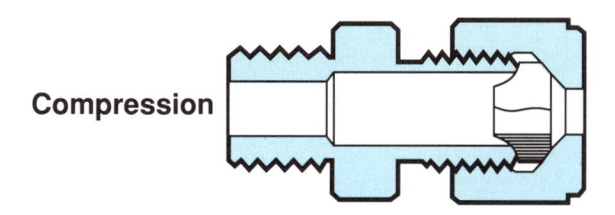

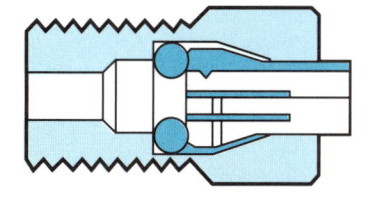

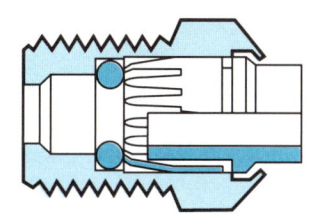

Barb

Compression

Slotted Collet Push-in

Multi-tooth push-in

Use:
To provide plumbing connection between pneumatic components such as valves and cylinders, and a compressed air source.

Operation:
Pneumatic automation systems depend on a network of pipe, tubing or hose, and fittings to transmit required compressed air energy.

Available Configurations:
- Hose and tubing can be made of buna-N rubber, thermoplastics, nylon, polyurethane, polyethylene and copper.
- Plastic tubing comes in varied colors and in straight, recoil braided and bonded variations.
- Fittings typically fall into three categories: barb, compression and push-in. They can also be used with or without clamps or crimped ferrules.
- Most pneumatic fittings have tapered pipe threads in sizes 10-32 through 2 inches.
- Two major types of push-in fittings use O-ring seals while other fittings depend on barbs and ferrules for sealing.

Sizing/Selection:
Sizing and selection for pipe and hose depend on pressure availability, flow and air reservoir capacity. Most systems are piped to the same size as component porting. When speed and accuracy are critical, consideration must be given to cumulative pressure drop resulting from turns, bends and orifice restriction in fittings. Push-in fittings usually provide full flow.

Application Notes:
- Fittings are now available in many extended forms such as swivel, elbow, tees, union, bulkhead, banjo tee, Y-shaped, and manifolds.
- Polyurethane tubing offers great flexibility and abrasive resistance within limited pressures and temperatures.
- Nylon tubing is resistant to many chemicals and liquids.
- Push-in fittings may cut up to 70% of assembly labor costs.
- Recoil hose assemblies of nylon and polyurethane are widely used with air tools.
- Teflon tape on these components is recommended for preventing leaking connections.
- Hose and tubing should be larger for longer runs to prevent pressure loss. Check the manufacturer's flow charts for recommended sizes.

Pneumatic Components

Grippers

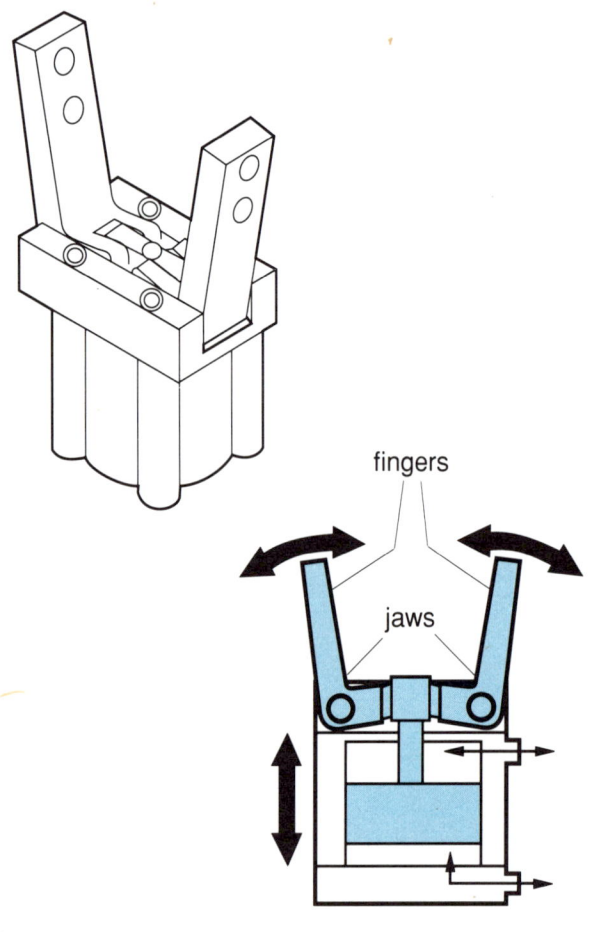

Available Configurations:
- Angular, whose jaws operate from a fixed pivot.
- Wide-opening angular, whose jaws open 180 degrees for easier part clearance.
- Parallel, whose jaws remain at a fixed angle but achieve their clamping action by opening and closing laterally.
- "Pseudo-parallel," whose jaws move in a largely parallel motion along an extremely shallow arc.
- Miniature, with very low force grippers in sizes up to 2" wide.
- Two-jaw, offering opposing gripping.
- Three-jaw, offering a centering gripping force.
- With embedded sensors used for determining whether the jaws are opened or closed a specified amount.

Sizing/Selection:
The maximum load that a gripper will hold will vary based upon the size of the part being picked up, the shape and texture of the part, the speed at which the part is being moved, the working pressure entering the gripper and the shape and length of the jaws. Selection criteria include: grip force needed, total jaw travel, internal and external grip size, operating pressure minimums and maximums, weight and overall size of gripper, operating environment and life cycle. In general, gripping forces will be proportional to the system pressure applied to the device.

Manufacturers' tables will assume some typical or average conditions such as jaw length, operating pressure and jaw travel. Many manufacturers offer specialized versions of their grippers which can exceed standard specifications through use of design alterations and alternate material selections.

Use:
For holding or clamping, typically in an assembly or loading/unloading operation.

Operation:
A pneumatic-based gripper typically relies on an internal cylinder and piston which is mechanically coupled to "jaws." These jaws mimic the finger action achieved by the human hand in grasping an object, but can also expand to grip from the inside. While a great many standard gripper bodies are offered, in most cases the gripper jaw tooling (or fingers) must be machined by the user to accommodate specific part or material handling.

Calibration/Adjustment:
Clamping force and speed are adjusted by the amount of air pressure and flow metered to the gripper. Jaw travel is typically determined by gripper and tooling design.

Pneumatic Components

Application Notes:

- It is best to "lock on" the object to be gripped by tooling the gripper fingers to conform to the object's shape or texture. For those applications where only friction is used to adhere to the object, the gripper should be sized larger.

- By combining sensors with grippers, confirmation of parts' grasp and even part sizing can be communicated as the gripper operates.

- Grippers can be ganged to grasp or support extremely long or heavy parts.

- Parts which vary in size may be grasped best by a parallel motion gripper since this configuration will maintain a consistent gripping area.

Index Tables

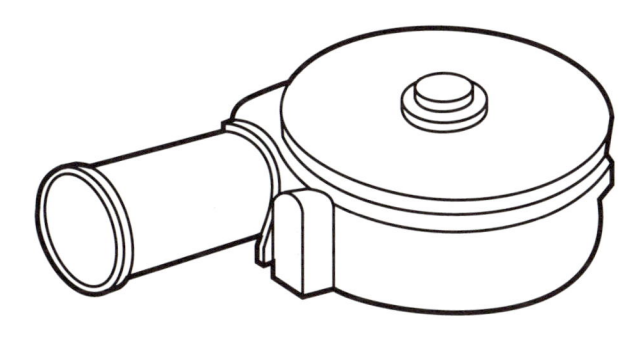

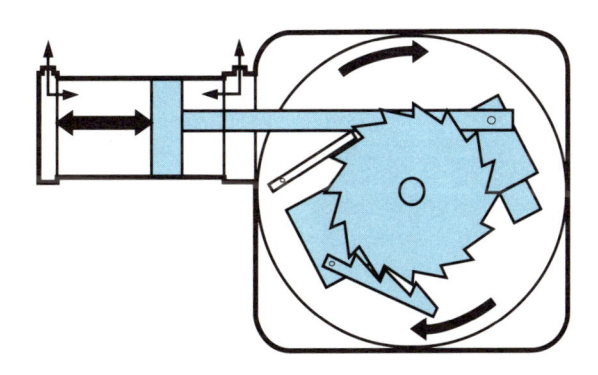

Operation:
A typical rotary design employs compressed air entering a cylinder with a ratchet pawl or rack gear on the end of its piston rod. This moves a ratchet or drive gear a specified distance, transforming linear into rotary motion, and rotating the table. Internal stops are usually provided to ensure the table will rotate the same distance each time. On its return stroke, the ratchet pawl slips over the teeth on the ratchet gear. In the case of the rack and drive gear rotary design, the drive gear is disengaged from the rack on the return stroke.

Available Configurations:
- Adjustable, which allows variance of cylinder stroke length for different graduations of table movement.
- Lift and turn, which moves the table vertically as well as rotationally.
- Internal solenoid valve, for engaging/disengaging the cylinder.
- Rotary or linear, for application suitability.

Sizing/Selection:
Selection of index tables is typically based upon the number of stations at which the table will be required to stop, the table base size, the weight of the part transported, cycle life and the turntable or linear table size.

Calibration/Adjustment:
Flow controls usually control the advancing speed of the table. Dampeners can also be adjusted to compensate for the starting and stopping shocks caused by the cylinder controlling the index table.

Use:
To provide precise, incremental, repeatable rotation of a turntable or the back-and-forth movement of a linear table. Both are typically used for transporting a part from one station to another.

Pneumatic Components

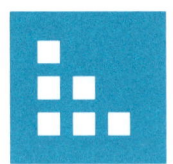

Application Notes:
- Use the smallest applicable size for the index table in order to maximize precision and cost effectiveness, and to minimize the effects of inertia.
- Because of their susceptibility to inertia, large diameter or heavily loaded rotary index tables should be fitted with dampers to compensate for shocks due to quick starting or stopping.

Presses

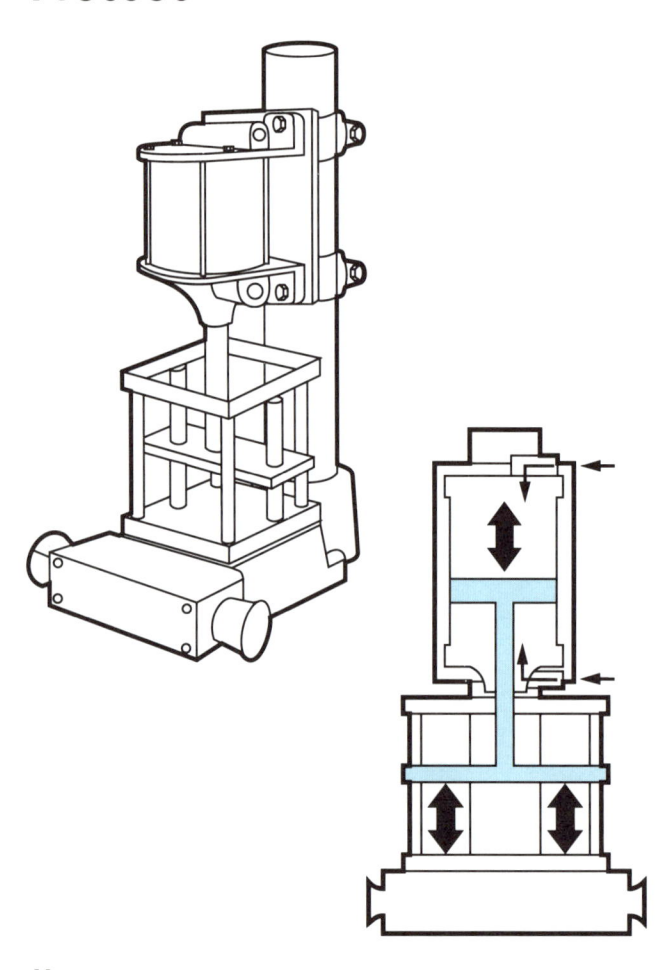

Use:
Provides a variety of holding, assembling or punching functions such as crimping, bending, forming, swagging, riveting and burnishing.

Operation:
Essentially one of a class of applications for air cylinders, presses typically utilize a cylinder mounted vertically on a stand or arbor. The up or down motion of the cylinder's piston provides the work. Pneumatic presses typically belong in the light duty category, developing forces of ten tons or less.

Available Configurations:
- Single-acting – provides force in one direction.
- Double-acting – provides force for extension and retraction.
- Arbor mount – often for heavy duty applications requiring milled tables.
- Column mount – for adjustable cylinder position.
- Double rod extending from both ends – minimizes piston rod deflection and allows stroke adjustment via external spacers.
- Booster or intensifier style – uses a large area piston pushing on a smaller area to increase force.
- Hand or remote actuation is often available.

Sizing/Selection:
Press sizing will be primarily determined by the force required, the physical dimensions of the part being worked and the production process. Most manufacturers will offer a choice of cylinders, thereby allowing changes in stroke length, press force, operating pressure, life expectancy and operating environment (i.e., temperature, moisture, maintenance level).

Calibration/Adjustment:
As with cylinders, pneumatic presses can be equipped with flow controls that allow metering of the flow of compressed air to change speed. Many also contain internal end stops and spacers which can be varied to control stroke length. Regulator valves can also be employed to reduce the force exerted by the press.

Pneumatic Components

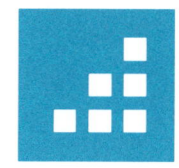

> **Application Notes:**
> - Presses should be securely mounted, with those operating at high speeds or pressures requiring properly reinforced mounting tables or bases.
> - The correct fixturing should always be employed to ensure that work is held in the correct position and in a secure manner.
> - The appropriate guards and interlock switches should be added. Two-hand safety switches must be used for manual operation.
> - Loading and unloading paths should be considered prior to press mounting and fixturing.

Programmable Logic Controllers

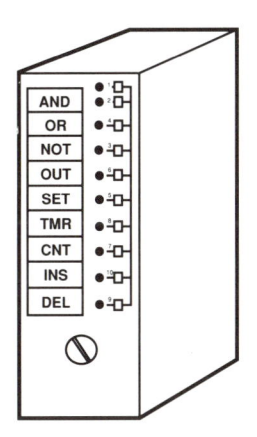

Use:
To turn various devices on and off, in specific sequences, at specific times in response to signals from sensors, switches, an operator, or other equipment.

Operation:
Programmable logic controllers (PLC's) employ microprocessor-based electronic circuits to make decisions as to what step in a particular automation application should occur at a particular time. The controllers depend upon a simplified programming process often referred to as *ladder logic*, which is set up by the user. This logic process allows the controller to make choices as to which solenoid valves, conveyors, vacuum pumps or other automation equipment should be turned on or off based upon whether a particular process has been successfully completed. The controller sequentially performs the program's logic, accepting input signals from sensors and switches, and acting upon them by energizing an output until the program is completed, whereupon the process is repeated. A key feature of programmable controllers is the versatility of their programming sequences and time durations. This is in contrast to more dedicated mechanical-pneumatic timing circuits and electrical relays which are typically configured for only one type of operation.

Available Configurations:
Programmable controllers vary, each having features which make them more or less desirable for a particular application. Most are geared to operate in the industrial environment. Some offer the ability to transport their programs to other programmable controllers. Most programmable controllers are designed to "communicate" their signals with other devices including numerically controlled machine tools, computers and other programmable controllers. There are various schemes for program storage among the controllers, with trade-offs existing between ease of program change and protection from program loss due to power failures. Many of these devices are equipped with special circuitry to protect the controller against extreme power fluctuations or spikes that are so prevalent in a factory environment. Many offer remote control capabilities for starting, stopping and reprogramming. Some controllers are expandable, offering the ability to increase control capabilities at a later time.

Sizing/Selection:
Programmable controllers are selected on the basis of: the number of items to be controlled (outputs); the number of incoming signals it will accept (inputs); its current carrying and switching capacity; the amount of internal memory it contains (which affects both the speed at which it can operate and the complexity of program it will accept); and the programming approach, or language, it utilizes. PLC's should also be evaluated for the life expectancy of the application.

Pneumatic Components

Calibration/Adjustment:
While the programming approach for each controller may vary, most tend to rely on a reference number to identify each step. In addition, most will further identify a circuit and the various devices which comprise the circuit. The functions performed typically include switching, timing, latching, counting and sequencing. The program logic is based upon performing a function in response to: a particular instruction or input (known as an *if/then function*); two or more instructions or inputs (known as an *if/and function*); one or another instruction or input (known as an *if/or function*); or not performing the function in response to a specific instruction or input (known as an *if/not function*).

Application Notes:
- Often, the same type of PLC is used throughout a location to reduce programming/reprogramming time, to make transference of programs from one PLC to another easier and to facilitate PLC service and repair.
- Most PLC's are available with a varying number of AC and/or DC voltage inputs and outputs.
- Often a PLC will allow the sequence of a machine's operation to change without changing any wiring.
- In addition to electronic programmable logic controllers, pneumatic logic may be considered.

Rotary Actuators

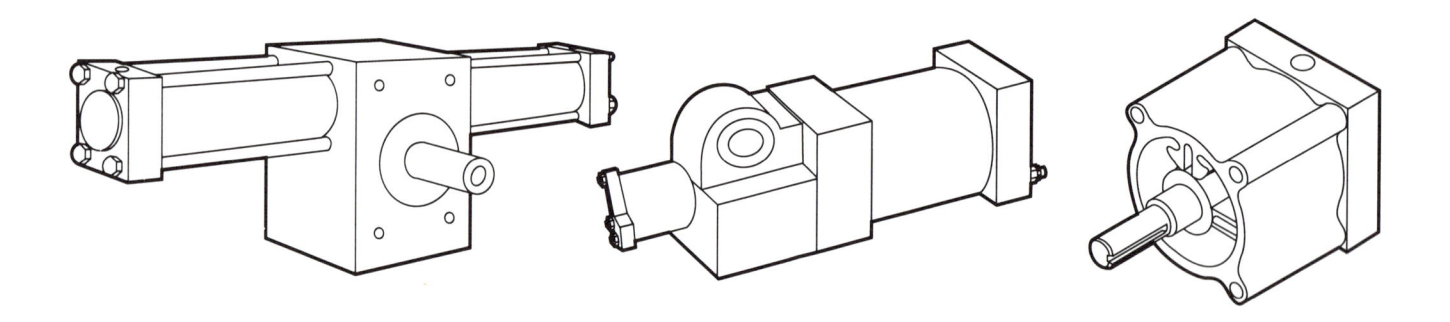

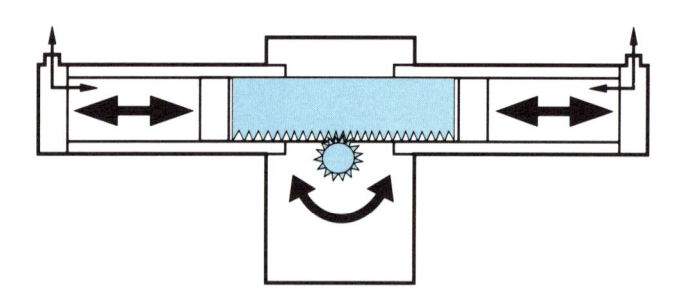

Use:
To provide reciprocating rotary movement, typically in the range of 0 to 360 degrees of rotation. Specialized designs offer more revolutions plus positive stopping points at various positions.

Operation:
The two most common types of rotary actuators are piston-driven and vane style. Piston-driven rotary actuators typically employ one or two pistons which are connected to a mechanical coupling. This coupling converts the linear piston movement into a rotary motion. The piston(s) function the same way they do in cylinders. Mechanical coupling techniques include rack-and-pinion, ratchet, helical spline, crankshaft, scotch yoke and piston-chain.

The second design, referred to as a vane style rotary actuator, consists of a vane connected to a shaft which is assembled into a housing. Pressurizing one side of a

Pneumatic Components

vane causes rotation of a shaft; applying pressure to the opposite side causes opposite rotation. A barrier, called a stator, isolates the two pressure chambers formed by this design.

Available Configurations:

- Single or double position piston-driven, with positive stops at end of strokes.
- Multi-position piston-driven, with up to five shaft stop positions.
- Miniature piston-driven, usually five inches or less in length.
- Tandem piston-driven, using two cylinders—or four pistons—for extra force.
- Multi-motion piston-driven – essentially a combination of a rotary actuator and a cylinder to achieve a combination rotary/linear motion.
- Unidirectional piston-driven – typically utilizing a ratcheting clutch to restrict output shaft motion to one direction for use in applications such as indexing.
- Air-oil – using oil bypass circuits pushed by air to achieve smooth shaft motion.
- Single vane models, capable of wider rotation, typically up to 280 degrees.
- Double vane models, which double the output torque, but are limited to a rotation of 100 degrees because of the second vane contacting the stator.

Sizing/Selection:

Determining factors in the selection of a rotary actuator include: torque output, speed, required degrees of rotation, number and points of positive stops, operating pressure, physical size, weight, life cycle and operating environment. Most options pertaining to cylinder selection, such as internal stroke adjustment, cushioning, and end-of-stroke sensors also apply to rotary actuator selection. Potential output shaft overloading is also a consideration. Most manufacturers offer a variety of shaft configurations and shaft adapters for attaching the actuator to other tooling. Multi-position rotary actuators are sometimes offered with customer specified stopping points. Checking manufacturer tables will yield information for torque output, tolerances, speeds and other selection criteria.

Calibration/Adjustment:

Rotary actuators can be equipped with flow controls that allow metering of the compressed air to change speed. End-of-rotation stopping points for piston-driven styles can be adjusted via internal cylinder stroke adjustments. Many also contain internal cylinder end stops and spacers which can control stroke length. Torque may be increased by maintaining a higher system pressure, using a larger sized rotary actuator or selecting a double piston or double vane unit.

Application Notes:

- Some rotary actuators are pre-lubricated for maintenance-free operation. Methods for non-lubricated service include special seals, coatings on the cylinder or housing walls, internal wicks or coated internal surfaces.

- The force of a rotary actuator can be boosted by increasing the inlet pressure to within the manufacturer's recommended limits.

- Over-center rotation can be controlled by employing self-contained air-over oil systems which cushion the actuator's motion and increase control.

- Cushions or bumpers can be utilized at the end of stroke positions in piston-driven actuators to help decelerate loads. Higher velocities and greater loads may require shock absorbers and/or external stops.

- Most rotary actuators have limitations as to the amount of axial load they can withstand, depending upon bearing type.

- Mid- or end-of-rotation position sensing can be accomplished through the use of reed or Hall effect switches, or other sensors.

Pneumatic Components

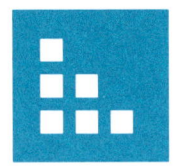

Sensors

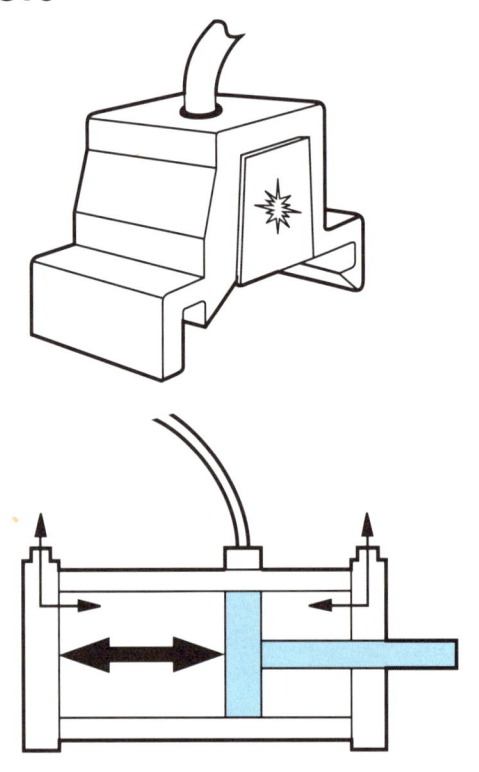

Use:
Typically used in applications to sense the position of an actuator indicating whether a work cycle has been successfully completed.

Operation:
Sensors used in pneumatic automation are based upon varying technologies such as magnetic, pneumatic, electronic, electrical and mechanical. The two main categories used to distinguish sensors are proximity and position. Proximity sensors serve as stroke completion sensors for cylinders and other actuators. Position sensors indicate either single or infinite positions along an actuator's travel path. Technology is now available which permits precise near end-of-stroke position sensing.

Sensors can be further subdivided into those that make contact with a device, such as a simple trip switch, reed switch, or potentiometer; and those that are non-contacting, such as Hall effect, or inductive proximity sensors.

Those sensors that are able to indicate relative position sometimes use conditioning circuitry to interpret their output. For example, a Hall effect sensor is usually coupled with an electronic circuit. This circuit can interpret the intensity of the sensor's signal, and indicate relative position of an actuator or part. Potentiometers, however, emit a signal of graduated intensity based upon increasing or decreasing resistance at a given point of their travel, and can be used to indicate relative position without extensive signal conditioning circuitry.

Many other, more specialized, varieties of sensors exist such as fiber optic sensors which use illumination to sense the presence of a part or actuator, and pressure differential switches which compare pressure in two parts of a pneumatic circuit and emit a signal based on sensing equal or unequal pressure conditions.

Available Configurations:
- Sensors and switches are offered with a variety of mounting means including: being banded around repairable and non-repairable cylinders, fixed to cylinder tie rods, mounted directly onto the cylinder end covers, and threaded into cylinder ports. They can also be mounted on other application moving parts.
- Pneumatic sensors typically mount directly in cylinder ports to provide an air signal when the piston motion stops. Their function is based on the difference in pressure on one or both sides of the piston.
- Trip switches are momentary contact devices operated by a lever or cam which is moved when a part or actuator reaches a certain position.
- Reed switches rely on a magnet placed on a moving part to bring two contacts together when the part passes near the switch.
- Hall effect switches and sensors also rely on magnets to change the direction of current flow across a sensing element when the magnetic field is close, with the output being proportional to the distance from the magnet.
- Inductive proximity switches sense a change in the electromagnetic or radio frequency signals emitted by the sensor when a piece of ferrous metal is near.
- Limit switches sense end of stroke positions in various actuator types.
- Position transducers are position sensors which exhibit linear voltage changes to indicate relative position.
- Encoders are pulse counters used to indicate position of a rotating shaft such as a stepper motor.

Pneumatic Components

Sizing/Selection:
A cylinder's port size will dictate a pneumatic sensor's sizing since most will operate across a wide range of system pressures. Electrical sensors will be sized based upon the electrical current they are capable of switching.

Calibration/Adjustment:
Most of the non-contacting electrical sensors need to be adjusted relative to the position between the sensor and the sensed element, whether it is a magnetic or ferrous material. Often, shims or spacers are used to adjust sensor position when sensors are inserted directly into cylinder end caps. However, some recent designs eliminate the need for shims and spacers, and utilize the same proximity switch probe length on both end covers of the cylinder. Trip or mechanically actuated switches are adjusted by placement relative to the contact mechanism. Pneumatic sensor operating pressures are usually preset at the factory.

Application Notes:
- Most sensors employing magnets have a maximum working temperature of 180 degrees to 200 degrees F. More specialized sensors are available from certain manufacturers.
- Reed and Hall effect switches are typically used to sense position anywhere during an actuator's stroke. Pressure switches generally are used to sense the actuator's end position.
- Sensors should be sized according to circuit current and voltage capacity. Individual component specifications should be reviewed before actual selection.
- Many sensors offer built-in protection against transient voltage spikes, reverse polarity or false pulses. If this protection is not built into the sensor's circuitry, it may be required elsewhere in the controlling circuitry.
- Many sensors come with a light emitting diode which activates when the sensor has closed the circuit. This feature can be extremely useful when debugging or servicing a pneumatics automation application.

Shock Absorbers, Linear Decelerators

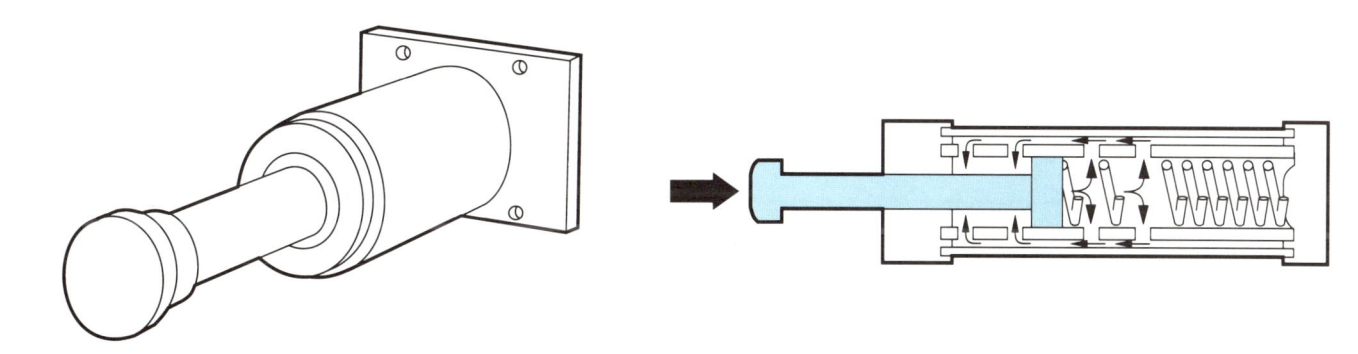

Use:
To minimize the collision effect between a moving body and a positive stop which, in turn, allows increased operating speed, decreased noise and increased equipment life.

Operation:
A variety of methods exist for achieving deceleration of moving items. Springs, rubber cushions, fluids (such as air or hydraulic oil) or other means can be employed which compress upon impact and absorb and dissipate kinetic energy. Each technique has certain characteristics in terms of performance, cost and longevity.

Pneumatic Components

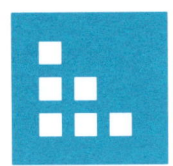

Available Configurations:

- Springs and rubber bumpers offer the advantage of simplicity and low cost. However, their stopping force is lowest at contact and this force increases until motion ceases. A large portion of the energy absorbed is returned to the item being cushioned after motion has ceased.
- Dashpots, (cylinders containing a fluid which is metered through a hole) absorb the largest amount of energy upon impact. Energy absorption levels decrease as motion ceases, returning little energy to the item being stopped.
- Cylinder cushions are typically located inside of a cylinder attached to its end caps. As with dashpots, they offer high initial energy absorption and high initial shock, which decreases as movement ceases.
- A linear decelerator-type shock absorber typically employs a fluid-filled double-walled cylinder. Upon impact, the cylinder's piston forces the fluid in its path through orifices in between the cylinder walls into the space behind the piston. This approach maintains an even resistance pressure throughout the entire deceleration period.
- A non-liquid shock absorber is filled with pellets that compress upon impact, lessening shock and any rebound effect. These devices must typically be replaced after a single use.

Sizing/Selection:

Selection of a suitable shock absorber depends upon the amount of energy that needs to be absorbed and the desired energy absorption pattern. This is a function of the weight of the moving load, the velocity at which the load is moving when it impacts the shock absorber, the forces that are propelling the load and the rate at which the motion cycles are occurring. Other considerations when selecting a shock absorber are: dimensional and weight limitations, cost, ability to withstand the operating environment and life expectancy.

Calibration/Adjustment:

Fluid-filled linear decelerators are either adjustable or non-adjustable. The adjustable type can be adjusted to a particular set of conditions so that the required stopping force is minimized. This is accomplished by varying the size of the fluid flow orifices.

There are two kinds of non-adjustable shock absorbers. The first is made to perform best at one set of conditions, and causes the lowest stopping force when subjected to those conditions. Typically, this is done by varying the fluid side wall orifice size or stroke length. Deviation from the specified conditions can cause large unwanted changes in the deceleration. The second kind is referred to by several names including self-compensating. Careful selection of the orifice sizes, spacing, fluids and other factors, causes the shock absorber to react so that it yields acceptable performance over a wide range of conditions.

Other types of shock absorbing methods can be adjusted only by placement, increasing or decreasing the number used, or by placing additional stops along the deceleration path.

Application Notes:

- Most shock absorbers are permanently sealed and do not require lubrication.
- "Side loading," the exertion of excessive pressure on an unsupported piston rod, can be a problem with shock absorbers just as it is with air cylinders. Fluid leakage, bearing and/or piston rod wear can result from exceeding the manufacturer's specifications.
- The accuracy of the automation application's characteristics affects the ability to correctly select the shock absorber. Carefully consider factors such as the kinetic energy generated by the application and the maximum energy that can be absorbed each hour without overheating the shock absorber.

Pneumatic Components

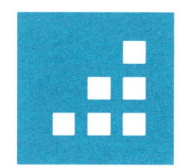

Silencers/Mufflers

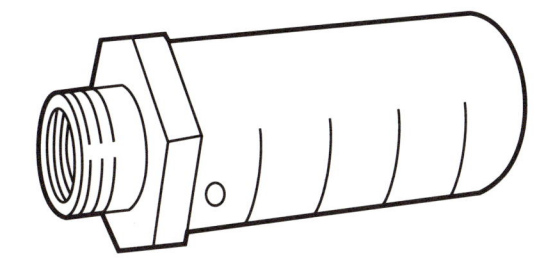

Use:
To reduce the noise levels caused by the rapid expelling of compressed air from a pneumatic device to the atmosphere.

Operation:
High velocity air escaping from a compressed air source creates shock waves as it mixes with static atmospheric air. A muffler or silencer provides a means to minimize these shock waves by reducing the shock wave formation and slowing the air's exit velocity as much as possible. Most muffling devices accomplish this by allowing the escaping high velocity air to expand, lowering its pressure and allowing it to mix with lower velocity air before exiting. One way this can be accomplished is by the introduction of baffles or porous material formed within or around an expansion chamber. Care must be taken in the design of a muffling device not to cause excessive back pressure which could impede the muffled device's operation.

Available Configurations:
- Porous filter-like devices made from plastic or metalized beads.
- A series of baffles and exhaust ports.
- Single or double expansion chambers.
- Reinforced foam rubber jackets.

Sizing/Selection
The goal of a muffling device is to bring the noise level of escaping compressed air to within a certain target requirement for human safety or comfort, without impeding pneumatic efficiency due to back pressure. This is a function of the exiting compressed air flow rate, the desired sound level at a given distance, the size of the exit port and the pressure drop through the muffling device. The manufacturers of these devices provide selection charts which plot the relationships between these parameters for given port sizes and sound attenuation levels.

Calibration/Adjustment:
Since these devices are produced in various sizes and configurations, little adjustment is necessary or possible.

Application Notes:
- These devices can be installed in any port where there is a rapid decompression of air.
- Silencers and mufflers may be prone to corrosion or clogging due to water, lubricants and dirt passing through them. Filtration may be required if this becomes a chronic problem.
- Icing may occur in a silencer or muffler. This is due to the component's internal surface temperature being reduced below atmospheric dew point because of a rapid expansion of the exhausted air as it passes through. In this case, double chamber mufflers or silencers can be used to graduate the reduction in pressure and reduce the likelihood of ice blockage.
- The Occupational Health and Safety Administration (OSHA) has set minimum safe noise levels. Their standards should be used in figuring pneumatic noise reduction levels.

Pneumatic Components

Vacuum Pumps

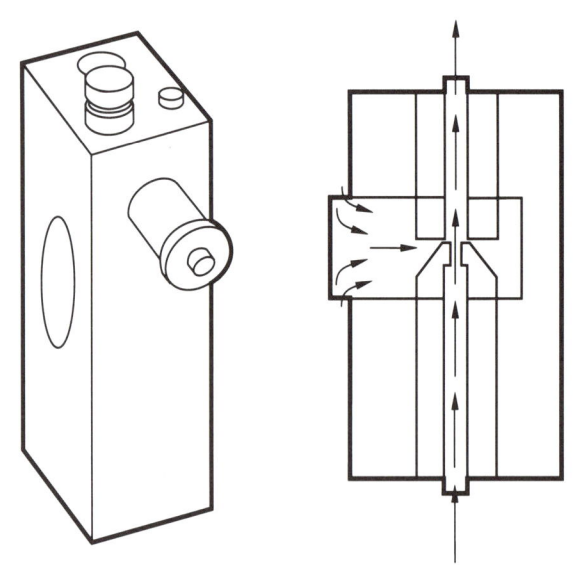

Use:
In conjunction with suction cups, used to grip objects which would be damaged or difficult to lift with conventional grippers.

Operation:
Two general methods for achieving vacuum exist: mechanical pumps, and ejectors which are powered by compressed air. The construction of mechanical pumps is similar to air compressors; they merely function in reverse. A piston, turbine or vane moves air from one area to another creating a vacuum. The difference between a mechanical vacuum pump and an air compressor is that air is released to the atmosphere instead of being compressed. In a compressed air-driven ejector, the vacuum is achieved by passing the pressurized air through a chamber via a small orifice or venturi. This chamber is sometimes connected to a suction cup or other contact device. As the compressed air moves through the venturi located in this chamber, it causes a difference in pressure between the air inside and outside of the chamber. This pressure differential causes a vacuum to be formed by an object obstructing the chamber's opening, such as when a part is picked up by a vacuum cup.

Available Configurations:
- Reciprocating piston, rotary vane or turbine designs are the most common mechanical types.
- Compressed air-driven, single-chamber ejector designs.
- Compressed air-driven, multi-chamber ejector, which generate more vacuum per the amount of compressed air consumed.

Sizing/Selection:
Selection of a vacuum pump and appropriate suction cup depends upon the level of vacuum required, the approximate volume of space which must be evacuated, how quickly the vacuum has to be produced and the distance between the vacuum pump and the point of evacuation. It is generally more efficient to place the vacuum pump as close to the point of evacuation as possible to reduce the volume of space required to be evacuated. This includes tubing, connectors and other conveyances. Also, utilizing a larger area of suction will result in more holding force with less vacuum required.

Calibration/Adjustment:
Mechanical vacuum pumps are adjusted by their operational rate. Compressed air vacuum pumps adjust vacuum levels via the amount of air pressure entering the pump. Normal shop air pressure is needed for optimum performance.

Application Notes:
- The more leakage likely in the application, the more capacity needed from the vacuum pump or ejector. Typically porous materials such as paper or cloth will cause more leakage when exposed to a vacuum than non-porous materials such as plastic or glass. Applying suction to coarse surfaces may increase leakage also.

- Vacuum force decreases with altitude since there is less barometric pressure to push the object being lifted against the point of vacuum. Calculations for the amount of vacuum required must be adjusted for height above sea level.

- Suction cup material should be selected according to environmental characteristics as well as size and shape. Temperature, contact with aggressive solutions and amount of surface friction are some of these factors.

- Proper filtration should be provided in order to ensure vacuum pump and generator operation.

Pneumatic Components

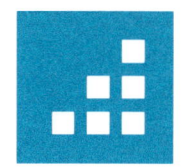

Valves, Directional Control

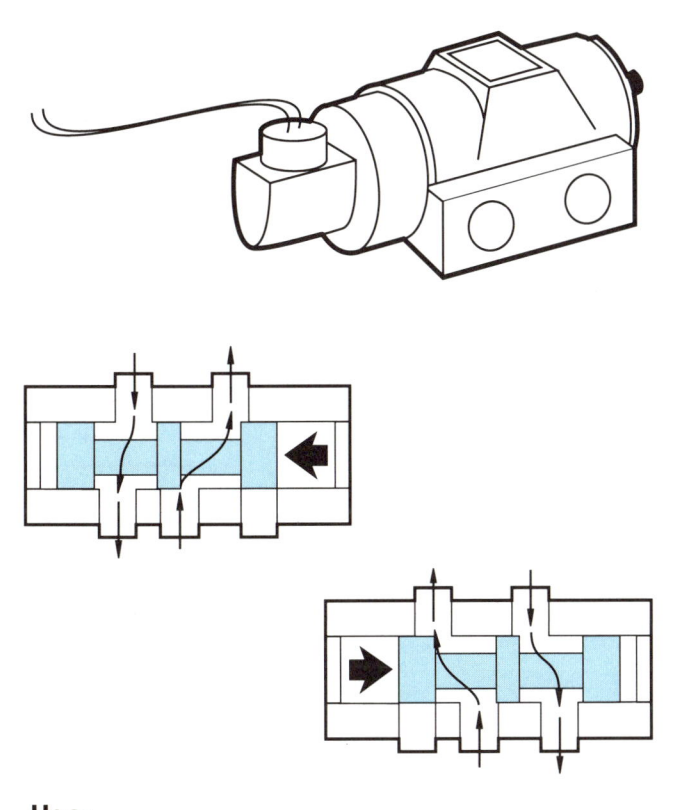

Use:
Controls the direction of compressed air to an actuator or general circuitry in a pneumatic system.

Operation:
A directional control valve typically consists of movable internal elements which open or close passageways. Directional control valves can be categorized by the number of inlets/outlets (also known as ports), and by the number of positions to which their internal parts are able to move. Hence, a two-position, four-way valve will move its internal element(s) to one of two positions, connecting its four inlet/outlet ports in a particular flow path. Directional valves also are distinguished by their internal design. The most common types are spool, poppet, slide and rotary disc. The means of activating the valve's movable element(s) can also distinguish a valve. The three most common are manual, electromagnetic and remote pneumatic pilot signal.

Available Configurations:
- Spool valves, which use a spool-shaped, internal, movable element to open and close passageways, allowing or stopping air flow between the ports.
- Poppet valves, which use one or more single plugs, or poppets, to open or close a port.
- Slide valves, which use a plate sliding back and forth to open or close the ports in either a lateral or rotary motion.
- Manual operation, using a lever, foot pedal, knob, handle or button actuated by a human; or mechanically by a roller, trip, plunger or toggle lever actuated by contact with another mechanical device.
- Electromagnetic, which typically employs a solenoid to move the internal elements, either directly or by pilot air. These same mechanisms may use internally directed pilot air, a spring or another solenoid to return the valve to its original position.
- Pneumatic, or remote air pilot, using air pressure from a remote source to move the valve's internal element(s), and using a spring or another air pilot to return the valve to its original position.

Sizing/Selection:
In the past, a directional control valve was sized by matching the valve's port size with the actuator's port size. This has often been found to be less economical than simply determining the amount of air used by the actuator in the circuit and comparing it to performance graphs available from the valve manufacturer. Selection formulas are also available, based upon rating parameters of the valve and the system's requirements.

Calibration/Adjustment:
Solenoid pilot and remote pilot valves may not function below a minimum operating pressure. Electrical solenoids also require that the supply voltage be maintained in a range that is typically +10% to -15% of the valve's rating.

Pneumatic Components

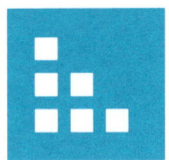

Application Notes:

- Caution: All supply air and electricity should be shut off before servicing a pneumatic valve or system. It is also necessary to depressurize all valve connecting lines.

- Some valves operate longer if a filter and an airline lubricator are installed upstream. Each valve manufacturer's instructions should be checked. Use manufacturer-specified oil.

- Valves may be installed as either in-line, with ports located directly in the valve body or with ports mounted in a separate base (manifold mount) to which the valve body is then attached. In-line valves occupy less space and are less expensive. Base mounted (manifold mount) valves provide mounting and installation convenience as well as simple, quick replacement. Valves also are provided in manifold or stacking designs that permit one common inlet to supply a series of valves.

- "Sandwich" flow controls and pressure regulators are two common accessories which fit between the base and body of some base mounted valves. (See *Valves, Flow Control,* below, and *Valves, Pressure Regulator,* pages 50-51.) There are 2 types of flow control valves: those installed in a pressure line, and those installed in an exhaust port.

- Flow controls are often mounted in the exhaust ports of spool type directional control valves to meter exhaust air from an actuator. (These are generally not suited for poppet valves.)

- Remote air pilot valves will discharge a pilot exhaust. If this is objectionable, the valves may often be modified to internally direct this exhaust to the main exhaust port. External piping may also be used to accomplish this.

- Low pressure (below the rated minimum shift pressure) and vacuum operation are sometimes possible with design modifications. Typically, these changes will require an external pilot supply pressure of a higher level for solenoid-operated valves.

- Electrical solenoid valves often contain a light which indicates that power has been delivered to the solenoid. This is an aid when troubleshooting a control circuit.

Valves, Flow Control

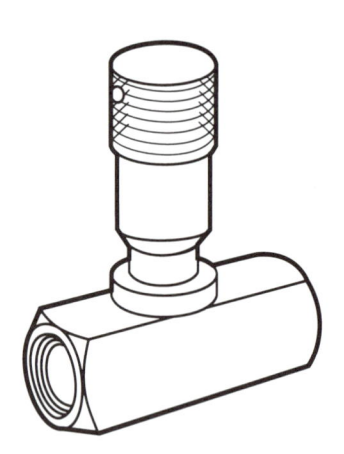

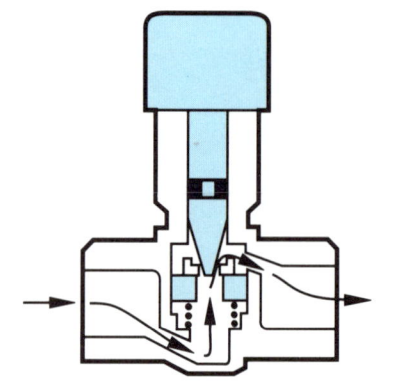

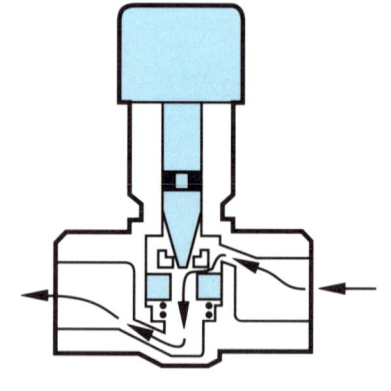

Use:
To control the flow of air; often to regulate the motion of an actuator.

Operation:
A flow control valve acts as a restriction in an air line. It can be located between an air source and an actuator, metering the air going in. Or, it can be located between an actuator and the atmosphere, metering the air going

Pneumatic Components

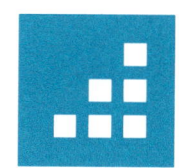

out. The greater the restriction, the less the flow rate. Flow rate is also affected by the operating pressure and temperature of the air in the system.

Available Configurations:
- Fixed orifice, which utilizes a preset size of restriction.
- Adjustable orifice, which utilizes some type of movable element to vary the size of the restriction.
- Globe valve – a coarse form of adjustable orifice flow control which utilizes a plug or poppet connected to a threaded shaft to meter the air flow.
- Needle valve – a more accurate form of adjustable orifice which utilizes a taper on the end of a threaded shaft to meter the air flow.
- Flow control with check valve – a very common configuration in pneumatic systems which allows air to be metered in one direction, and to flow nearly unrestricted in the opposite direction.
- In-line, mounted in the hose or tubing supplying air to an actuator.
- In-port, threaded into the intake or exhaust port of an actuator

Sizing/Selection:
Flow control sizing is based upon the flow rate passing through the valve for a given inlet pressure, its ratio-to-exhaust and the position of adjustment, if any. They are also sized to fit specific port or hose diameters. Some will control the air flow in only one direction; some will control it both ways. Generally for large flow applications, flow control valves should be sized above the component pipe diameter due to the small orifice size in most flow controls.

Calibration/Adjustment:
Fixed flow controls can be calibrated against a standard orifice. This calibration is for the amount of flow at defined conditions such as pressures exerted at the inlet or outlet, temperature and moisture. Adjustable flow controls often have indicating devices which can be used for reference while adjusting the system. These reference points are useful for later readjustment. Also, most flow controls are selected and then adjusted on the basis of steady conditions. Since pneumatic systems experience variations, especially during start-up periods, the full range of system operating conditions should be observed. System variations at start-up are often related to temperature. Until a system stabilizes, flow controls will exhibit varying performance due to density changes in the air, parts contraction and possible icing. Exposure to excessive heat can change flow rates, particularly lesser flow rates.

Application Notes:
- Even though the port size of an actuator or the tubing size being used often determines the flow control size, the flow coefficient should also be considered. This will help to conserve air, reduce cost and maximize results.

- When placement of the flow control in the port of the actuator is not possible, in-line or base-mounted flow controls can be used.

- If flow controls are located close to a moving actuator casing, the risk of kinking the inlet tubing increases. To prevent kinking, flow controls which swivel can be used. This eliminates the twisting of the tubing which can occur near the flow control's connection points.

- Flow controls should be chosen which minimize the number of connections. This, in turn, minimizes potential leakage points. Flow controls are now available with a factory-applied thread sealant which can minimize assembly time and promote leak-free joints.

- Many flow controls are equipped with lock nuts which, once tightened, prevent the flow control's needle valve stem from being moved.

- Cleanliness is a very important requirement for flow control valves. Since the valve is a restricting device, contamination can occur as foreign materials and moisture pass through. Air filtration and drying are recommended remedies to this problem. For small flows, a mesh strainer should be installed on both sides of the flow control valve to guard against changes in flow rates due to contamination.

Pneumatic Components

Valves, Pressure Regulator

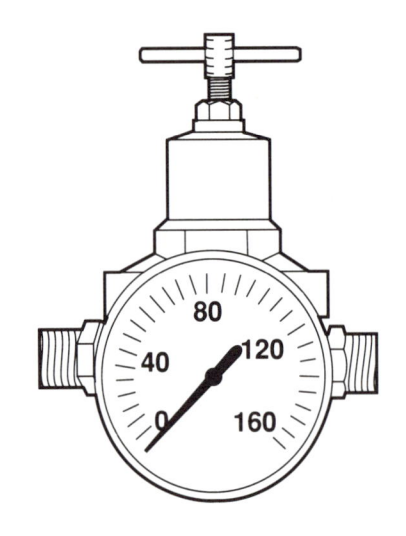

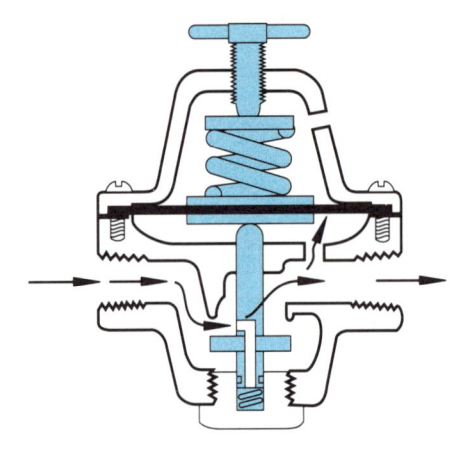

Use:
To reduce a higher level of air pressure to a lower level for component operation, and to maintain constant air pressure at the lower level as the air flow rate varies.

Operation:
Pressure regulators sense outlet pressure and adjust an internal movable element in order to maintain a constant outlet pressure. Generally, this is accomplished by allowing the outlet air flow to pressurize a piston or diaphragm. These movable elements are typically attached to a poppet-style valve that meters the inlet air flow. As outlet pressure increases or decreases, it presses on or releases the piston or diaphragm, causing the poppet valve to vary its opening and throttle the inlet air pressure. The level of outlet pressure desired can be changed by a screw which increases or decreases a spring force exerted against the opposite side of the piston or diaphragm.

Available Configurations:
- Piston type, in which the outlet pressure is sensed by a piston.
- Diaphragm type, in which the outlet pressure is sensed by a diaphragm.
- Venting, which provides a small amount of outlet pressure relief by opening an exhaust hole if the outlet pressure exceeds a desired level.
- Non-venting, in which the regulator does not have the ability to exhaust air.
- Reverse flow regulators, which have the ability to pass air flow in the opposite direction when the inlet pressure is exhausted.
- Pilot controlled regulator, which uses a remote source of pressurized air to act against the piston or diaphragm instead of a spring. This arrangement requires a controlled pilot pressure to adjust the outlet pressure level. (This controlled pilot can be achieved through use of a low flow pilot regulator or with electronically controlled solenoid valves.)
- Balanced regulator, which is designed to be insensitive to variations in supply pressure.
- Unbalanced regulators, which are influenced by supply pressure (although in small regulators this influence may be negligible).

Sizing/Selection:
Regulators should be sized from performance specifications published by each manufacturer. Using the required pressure in the system and the flow rate, a regulator should be selected that has a forward flow curve which exceeds the amount required by the system. Flow rates are calculated from the volume of air used by an actuator, its operating pressures and its cycle time. One should always be certain that the inlet supply pressure does not exceed the chosen regulator's inlet rating.

Pneumatic Components

Calibration/Adjustment:
A regulator generally has a gauge port into which a pressure gauge is installed to monitor the downstream circuit pressure. With no flow occurring in the circuit, the regulator's adjustment screw or knob can be turned until the desired maximum system pressure is reached. For regulators controlled by remote pilot pressure, the adjustment procedure is performed on the pilot pressure regulator. Operate the system and observe any pressure variations. If further adjustments are required, turn the adjusting screw, knob or remote pilot adjuster until a satisfactory pressure level is reached.

CAUTION: Always approach the maximum pressure setting from a lower pressure. Hysteresis in the regulator mechanism can cause inaccurate pressure settings if adjusted by lowering the pressure.

Application Notes:
- Caution: Install most regulators "upstream" of directional control valves in order to keep the flow in only one direction. Flow on the outlet side of directional control valves often changes direction. Some regulators, designed for reverse flow, could be located "downstream" of directional control valves, but not all regulators have this capability.
- Regulators have several spring ranges that limit their outlet pressures. A high spring range can be used for a wider range of applications, but cannot be adjusted as precisely, nor will their output performance be as exact. Ideal spring range selections are rated about 20 percent to 40 percent above the desired outlet pressure.
- Piston regulators are effective where long-life requirements are called for, but they exhibit more hysteresis. Diaphragm regulators are more accurate but they are more sensitive to fatigue.

Valves, Specialty

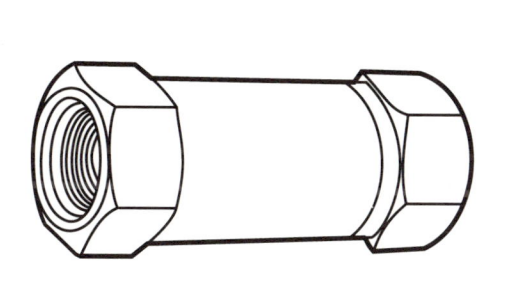

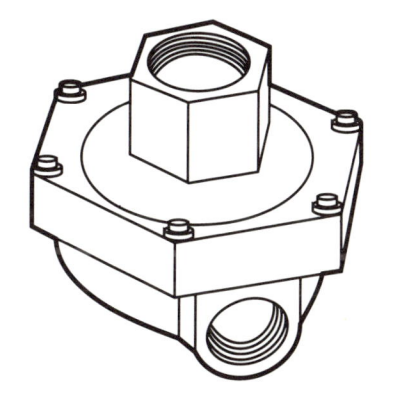

Check Valve **Quick Exhaust Valve** **Impulse Valve**

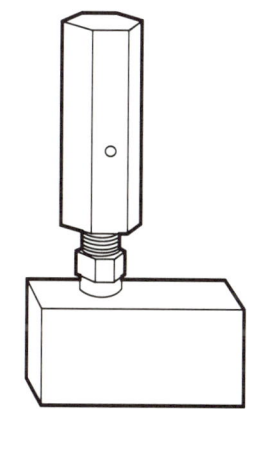

Use:
To provide a variety of specialized functions such as: restricting air flow to one direction, allowing air to exhaust rapidly from an actuator, providing a time delay, and serving as a momentary system pressure relief.

Pneumatic Components

Check Valve

(diagram labels: flow, no flow)

Shuttle Valve

Quick Exhaust Valve

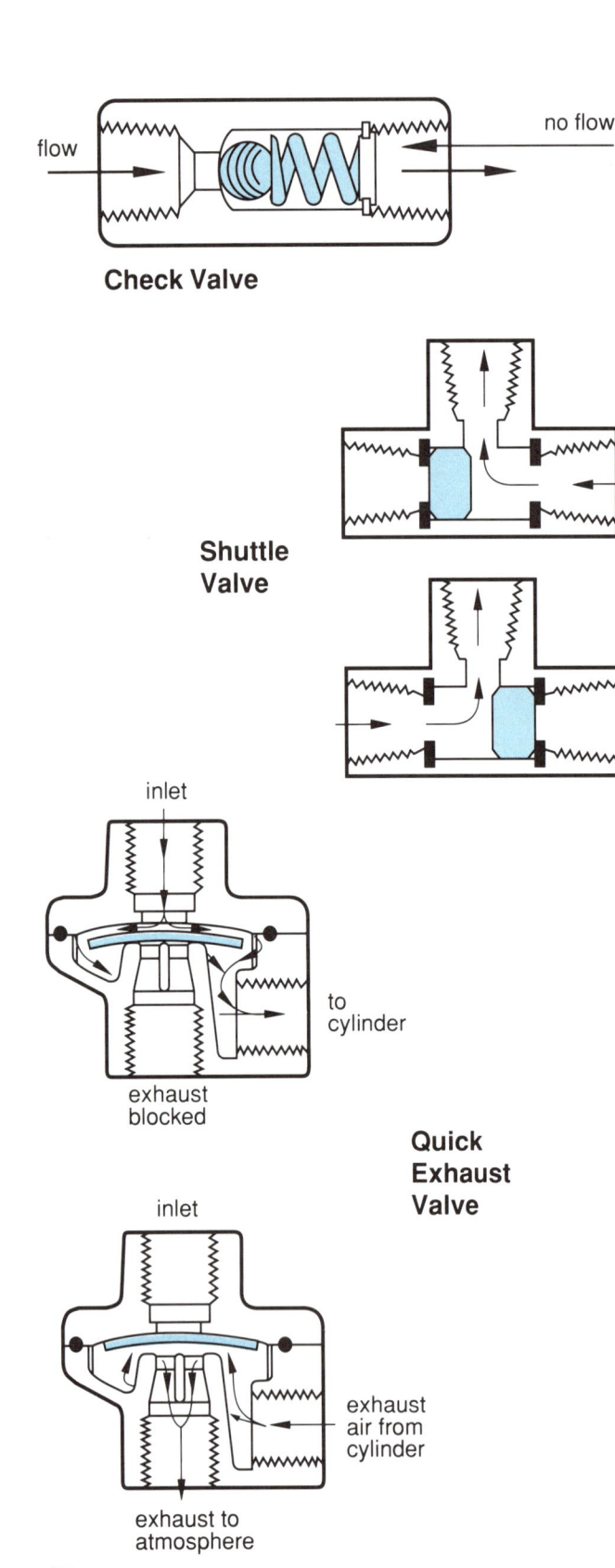

(diagram labels: inlet, to cylinder, exhaust blocked, inlet, exhaust air from cylinder, exhaust to atmosphere)

Operation:

There are times when only a one-way air flow is required. For example, it would sometimes be dangerous for a cylinder to be allowed to retract should compressed air be cutoff. **Check valves** usually depend upon a poppet held into a seat by a spring to provide this one-way flow condition. When air flows from the seat end of the valve, it easily overcomes the spring pushing on the poppet. But when air enters from the opposite end, it pushes the poppet onto its seat providing an effective seal.

Shuttle valves are automatic flow path selectors which allow the higher of two pressures to be directed into a flow path.

Time delay valves are one of the ways used to introduce a time delay into a pneumatic circuit, and are made in a variety of forms.

As their name implies, **quick exhaust** valves allow a cylinder to use the same port for rapid depressurization as for pressurization by opening the port to the atmosphere on the exhaust stroke. This action provides a one-way gatekeeping function for air flow similar to a check valve.

Impulse valves are used to momentarily interrupt a constant air flow to a pilot valve operating a directional valve, allowing the directional valve to change position. Impulse valves trigger by sensing back pressure in the system.

Available Configurations:

-Check Valves-
- A normally closed check valve is the most common form with ports on opposite ends of a tube-shaped body.

-Shuttle Valves-
- Shuttle valves provide a choice of flow from one inlet or another into a single outlet, driven by the pressure exerted from one or the other of these inlets.

-Time Delay Valves
- Piston/air timers resemble a standard pneumatic cylinder with a flow control and provide time delaying action by using the slowed movement of the piston to contact with a switch, valve or another actuator.

Pneumatic Components

- Some time delay valves depend upon the metering of compressed air into a reservoir at an adjustable rate until reaching a threshold level high enough to open a valve allowing the main air flow to pass.
- Mechanical escapement timers use an air operated piston to wind up a spring, then release a valve at the end of the spring's travel.

-Quick Exhaust Valves-
- Diaphragm-based quick exhaust valves generally depend on a rubber disc which pushes against either the inlet or exhaust portion of the valve, depending upon whether the actuator is filling or exhausting. This allows the actuator to push its exhausted compressed air directly into the atmosphere rather than pipe it back to a valve for exhausting.

-Impulse Valves-
- Typically used to control directional valves which are shifted by pilot air pressure, an impulse valve momentarily interrupts pilot pressure applied to one side of the directional valve. This allows the valve to shift when pressurized from the other direction. It is often used when the pilot air to one side of a directional valve cannot be automatically shut off, as when controlled by a manual valve.

Sizing/Selection:
Because of the special nature of check valves, a few body sizes are usually manufactured which can withstand a range of system pressures. Quick exhaust valves are typically placed in or near an actuator port and consequently are packaged to fit most standard port sizes such as 1/8", 1/4" and 3/8". For time delay valves, the timing range of a particular valve (which is usually in the one minute or less category) will be the determining factor. Typically the valves are offered in the same body with the same sized ports, but are rated for different delay periods. Impulse valves also are restricted to a few standard sizes but are usually offered to either fit a pressure or bleed type pilot directional valve.

Calibration/Adjustment:
Check valves, quick exhaust valves and impulse valves have no adjustment capability; however, check valves can be ordered with varying spring tension. Most timing valves are equipped with controls that allow variable metering of the air flow through the device to change its speed and force. Many also contain internal end stops and external spacers which can be varied to control stroke length.

Application Notes:
- Check valves are made in two general varieties: soft-seated, which offer 100 percent air shut off, and hard-seated, which tend to leak slowly over time. Soft-seated check valves will wear faster than hard-seated check valves.

- Shuttle valves can be used in an auxiliary or back-up circuit to maintain pressure by leading two independent pressure sources into each side of the valve. If one branch fails, the valve poppet will shuttle, closing off the depressurized branch and allowing the pressurized branch to flow freely.

- If possible, quick exhaust valves should be mounted in the port of the actuator to reduce exhaust time. Also, installing a larger diameter fitting between the actuator and the quick exhaust valve will lessen exhaust time for the same reason.

- Time delay valves, which meter smaller volumes of air, can become less accurate as time intervals increase. Greater accuracy will result by using valves with larger reservoirs and orifice diameters, even when smaller time intervals are required.

Where to Learn More About Pneumatics

Air Control Valves and Air Preparation Units. Cleveland: Parker Fluidpower, 1980.

Air Logic. Ralph Culbertson. Cleveland: Penton Publ., 1988.

Industrial Pneumatic Technology, Parker Fluidpower, 1980.

The Analysis and Design of Pneumatic Systems. Blaine W. Andersen. Melbourne, FL: Krieger, 1976.

Applied Hydraulics & Pneumatics in Industry. Trade and Technical Press Editors. Brookfield, VT: Brookfield Publ., 1980.

Basic Fluid Power. Dudley A. Pease. Englewood Cliffs, NJ: Prentice Hall, 1967.

Basics of Pneumatics. Videotape. Milwaukee, WI: National Fluid Power Association.

Communication, Including Graphic Symbols and Metric Units. Volume A, NFPA Encyclopedia of Fluid Power Standards. Milwaukee, WI: National Fluid Power Association, 1989.

Compressed Air and Gas Handbook, 4th ed. New York: Compressed Air and Gas Institute, 1973.

Cutting Costs with Pneumatics. Hauppauge, NY: Festo Corporation, 1988.

Design of Pneumatic and Fluidic Control Systems. Edward L. Holbrook. Milwaukie, OR: PECH Publishing, 1984.

Designing Pneumatic Control Circuits: Efficient Techniques for Practical Applications. B. E. McCord. New York: Marcel Dekker, 1983.

Dictionary of Hydraulics and Pneumatics. Gunter Neubert. New York: State Mutual Book, 1980.

Electrical Control of Fluid Power. Dallas: Womack Educational Publications.

Fluid Power Design Handbook. Frank Yeaple. New York: Marcel Dekker, 1990.

Fluid Power: Pneumatics. Olaf Johnson. Melbourne, FL: Krieger, 1975.

Fluid Power Handbook and Directory. Cleveland: Penton Publishing, 1989.

Fluid Power in Plant and Field. Dallas: Womack Educational Publications.

Where to Learn More About Pneumatics

Fundamentals of Control Technology. Hauppauge, NY: Festo Corporation, 1988.

Hydraulics and Pneumatics. Monthly periodical. Cleveland: Penton Publ.

Industrial Fluid Power, Volumes I, II and III. Dallas: Womack Educational Publications.

Industrial Pneumatics: Automation Using Compressed Air. Joucomatic/Techno-Nathan International. Paris: LaNouvelle Libraries, 1986.

Industrial Pneumatic Control. Z. J. Lansky and L. F. Schrader, Jr. New York: Marcel Dekker, 1986.

Industrial Pneumatic Technology. Cleveland: Parker Fluidpower, 1980.

Introduction to Pneumatics. Hauppauge, New York: Festo Corporation, 1989.

Machine Design. Annual fluid power issue. Cleveland: Penton Publishing.

Maintenance of Pneumatic Equipment & Systems. Hauppauge, NY: Festo Corporation, 1984.

Pneumatic Applications. Hauppauge, NY: Festo Corporation, 1986.

Pneumatic Circuits and Low Cost Automation. J. R. Fawcett. London: Brookfield Publishing Co., 1969.

Pneumatic Controls. Hauppauge, NY: Festo Corporation, 1987.

Pneumatic Handbook, 5th ed. Institute for Power Systems. New York: State Mutual Book, 1979.

Pneumatic Handbook, 6th ed. R. H. Warring. Houston, TX: Gulf Publishing, 1982.

Pneumatic Handbooks, 6th ed. Trade and Technical Press Editors. Brookfield, VT: Brookfield Publ., 1984.

Pneumatic Tools for Industrial Applications. The American Society of Tool and Manufacturing Engineers. Dearborn: The Association, 1965.

Pneumatics Explained (includes audiovisual package). Leo Rizzo. New York: Bergwall, 1984.

Robotics for Safety and Profit. Paul Jones. Chicago: P/P Publications, 1981.

Testing Fluid Power Components. Robert A. Nasca. New York: Industrial Press Inc., 1990.

Sources for Pneumatic Components

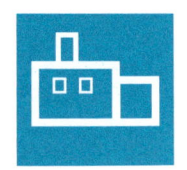

Air Collets/Chucks
FESTO Corporation
S-P Fluid Power

Air Motors
The ARO Corp.
Boston Gear Div./Imo Industries Inc.
Control Line Equipment, Inc.
Fenner Fluid Power
Gast Manufacturing Corporation
Horton Manufacturing Co., Inc.
Lynair, Inc.
Miller Fluid Power (linear)
TAIYO AMERICA, INC.

Air Tools
AAA Products International
The ARO Corp.
FABCO-AIR INC.
Legris Incorporated
Nycoil Company

Air Vises, Clamps
Peninsular, Inc.

Boosters/Intensifiers
Air-Dro, Div. of Decatur Cylinder Inc.
Atlas Cylinder Corp.
T. J. Brooks Company
Control Line Equipment, Inc.
FABCO-AIR INC.
Galland Henning Nopak, Inc.
Haskel, Inc.
Hennells Inc., Div. of C.M. Smillie
Hydro-Line Manufacturing Co.
Lynair, Inc.
M-C Industries Inc.
Miller Fluid Power
Milwaukee Cylinder Co.
Mosier Industries
Ortman Fluid Power
Parker Hannifin Corp.
Racine Fluid Power - Bosch Pneumatics
S-P Fluid Power
Schrader Bellows
The Sheffer Corporation
Vickers Incorporated

Cylinders
Advance Automation Co.
Air-Dro, Div. of Decatur Cylinder Inc.
The ARO Corp.
Atlas Cylinder Corp.
Bimba Manufacturing Co.
BOBALEE Hydraulics
Boston Gear Div./Imo Industries Inc.
T. J. Brooks Company
Control Line Equipment, Inc.
FABCO-AIR INC.
FESTO Corporation
Flairline Division/ACE Controls Inc.
Galland Henning Nopak, Inc.
Hanna Corp.
Hennells Inc., Div. of C. M. Smillie
Hunger U.S. Special Hydraulic Cylinders Corp.
Hydro-Line Manufacturing Co.
Kilsby-Roberts
Lynair, Inc.
M-C Industries Inc.
Mailhot Hydraulique Inc.
Mecman Inc.
Miller Fluid Power
Milwaukee Cylinder Co.
Mosier Industries
Origa Corporation
Ortman Fluid Power
Parker Hannifin Corp.
Peninsular Inc.
PHD, Inc.
Racine Fluid Power - Bosch Pneumatics
Rexroth Corp., Pneumatics Division
S-P Fluid Power
Schrader Bellows
The Sheffer Corporation
TAIYO AMERICA, INC.
Vickers Incorporated
Wilkerson Corporation

Dryers
Parker Hannifin Corp.
Wilkerson Corporation

Electropneumatic Positioners/Converters
Control Line Equipment, Inc.
FESTO Corporation (Interface devices)
Galland Henning Nopak, Inc. (Position reporting systems)
Hydro-Line Manufacturing Co. (Position sensing)
Miller Fluid Power (Interfaces)
Mosier Industries
Peninsular, Inc. (Position sensing)
Rexroth Corp., Pneumatics Division (Interface devices and positioners)

Filters/Regulators/Lubricators (FRL's)
ASCO Pneumatic Controls/ Joucomatic Div.
Amflo Products
The ARO Corp.
Automatic Switch Company
Boston Gear Div./Imo Industries Inc.
FESTO Corporation
Master Pneumatic-Detroit Inc.
Mecman Inc.
Miller Fluid Power
Parker Hannifin Corp.
Purolator Products Co., Facet Filter Products Division
Racine Fluid Power - Bosch Pneumatics
Rexroth Corp., Pneumatics Division
Ross Operating Valve Co.
Schrader Bellows
TAIYO AMERICA, INC.
Vickers Incorporated
Wilkerson Corporation

Sources for Pneumatic Components

Fittings/Connectors/Hose/Tubing/Adaptors
American Couplings Company
Amflo Products
Anderson Copper and Brass Company
The ARO Corp.
Chicago Fittings Corp.
FABCO-AIR INC.
FESTO Corporation
Fluid Product Sales Div. – Dana Corp
Furon Co. – Synflex Division
The Gates Rubber Co.
Hansen Coupling Division
Hofmann Engineering Co. Inc.
Imperial Eastman, Imperial Division
Legris Incorporated
M-C Industries Inc.
Mecman Inc.
Miller Fluid Power
Nycoil Company
Parker Hannifin Corp.
Perfecting Services, Inc.
Racine Fluid Power -
   Bosch Pneumatics
J. H. Roberts Industries Inc.
Schrader Bellows
Snap-Tite Inc.
Vickers Incorporated
World Wide Fittings Corp.

Grippers
FABCO-AIR INC.
Parker Hannifin Corp.
PHD, Inc.
TAIYO AMERICA, INC.
Teknocraft, Inc.
   (Sense and grip and electro-
   pneumatic vacuum grippers)

Index Tables
FESTO Corporation
Parker Hannifin Corp.
Schrader Bellows

Presses
FABCO-AIR INC.

Programmable Logic Controllers
Boston Gear Div./Imo Industries Inc.
FESTO Corporation
Galland Henning Nopak, Inc.
Miller Fluid Power
Mosier Industries
Ortman Fluid Power
Parker Hannifin Corp.
Peninsular, Inc. (Position sensing)
Racine Fluid Power -
   Bosch Pneumatics
Telemecanique, Inc.
Vickers Incorporated

Rotary Actuators
Barksdale Controls Div. and
   Boston Gear Div./Imo Industries Inc.
Bimba Manufacturing Co.
FESTO Corporation
Horton Manufacturing Co., Inc.
Hunger U.S. Special Hydraulic
   Cylinders Corp.
G. W. Lisk Company, Inc.
M-C Industries Inc.
Mecman Inc.
Micro-Precision Operations Inc.
Parker Hannifin Corp.
PHD, Inc.
Racine Fluid Power -
   Bosch Pneumatics
S-P Fluid Power
Schrader Bellows
TAIYO AMERICA, INC.
Vickers Incorporated

Sensors/Switches
Advance Automation
Barksdale Controls Div./
   Imo Industries Inc.
Bimba Manufacturing Co.
CEC Instruments Div./
   Imo Industries Inc.
Control Line Equipment, Inc.
Crouzet Corp.
FABCO-AIR INC.
FESTO Corporation
Hanna Corp.
Hedland Division,
   Racine Federated Inc.
Hydro-Line Manufacturing Co.
G. W. Lisk Company, Inc.
Mecman Inc.
Miller Fluid Power
Mosier Industries
PHD, Inc.
Parker Hannifin Corp.
Peninsular, Inc.
Racine Fluid Power -
   Bosch Pneumatics
Rexroth Corp., Pneumatics Division
Schrader Bellows
Sigma-Netics Inc.
TAIYO AMERICA, INC.
Teknocraft, Inc.
Telemecanique, Inc.
Vickers Incorporated

Shock Absorbers, Linear Decelerators
ACE Controls Inc.
FESTO Corporation
Hennells Inc., Div. of C.M. Smillie
   (Springs)
Mecman Inc.
Peninsular, Inc. (De-cel)
Schrader Bellows

Silencers/Mufflers
The ARO Corp.
Boston Gear Div./Imo Industries Inc.
FESTO Corporation
Master Pneumatic-Detroit Inc.
Mecman Inc.
Miller Fluid Power
Mosier Industries
Nycoil Company
Parker Hannifin Corp.
PIAB Vacuum Products
Racine Fluid Power -
   Bosch Pneumatics
Ross Operating Valve Co.
Schrader Bellows
TAIYO AMERICA, INC.
Versa Products Co., Inc.
Vickers Incorporated
Wilkerson Corporation

Sources for Pneumatic Components

Vacuum Pumps
FABCO-AIR INC.
FESTO Corporation
Gast Manufacturing Corporation
Mecman Inc.
PIAB Vacuum Products
Racine Fluid Power -
    Bosch Pneumatics
Teknocraft, Inc.

Valves - Directional Control
AAA Products International
ASCO Pneumatic Controls/
    Joucomatic Div.
The ARO Corp.
Automatic Switch Company
Barksdale Controls Div. and
    Boston Gear Div./Imo Industries Inc.
Bimba Manufacturing Co.
Control Line Equipment, Inc.
Crouzet Corp.
Deltrol Fluid Products
FABCO-AIR INC.
FESTO Corporation
Galland Henning Nopak, Inc.
Honeywell Inc., Skinner Valve Division
Kepner Products Company
Legris Incorporated (Manifolds)
Lexair, Inc.
G. W. Lisk Company, Inc.
Mecman Inc.
Miller Fluid Power
Moog Controls Inc.
Mosier Industries
Parker Hannifin Corp.
Racine Fluid Power -
    Bosch Pneumatics
Rexroth Corp., Pneumatics Division
Ross Operating Valve Co.
Schrader Bellows
Snap-Tite Inc.
TAIYO AMERICA, INC.
Teknocraft, Inc.
Telemecanique, Inc.
Valves Inc./
    Division of ACE Controls Inc.
Versa Products Co., Inc.
Vickers Incorporated

Valves - Flow Control
AAA Products International
ASCO Pneumatic Controls/
    Joucomatic Div.
The ARO Corp.
Automatic Switch Company
Bimba Manufacturing Co.
Boston Gear Div./Imo Industries Inc.
Control Line Equipment, Inc.
Crouzet Corp.
Deltrol Fluid Products
FABCO-AIR INC.
Fenner Fluid Power
FESTO Corporation
Galland Henning Nopak, Inc.
Honeywell Inc., Skinner Valve Division
Imperial Eastman, Imperial Division
Kepner Products Company
Legris Incorporated
G. W. Lisk Company, Inc.
M-C Industries Inc.
Mecman Inc.
Miller Fluid Power
Moog Controls Inc.
Mosier Industries
Nycoil Company
Parker Hannifin Corp.
Racine Fluid Power -
    Bosch Pneumatics
Rexroth Corp., Pneumatics Division
Ross Operating Valve Co.
Schrader Bellows
Snap-Tite Inc.
Telemecanique, Inc.
Valves Inc./
    Division of ACE Controls Inc.
Versa Products Co., Inc.
Vickers Incorporated

Valves - Pressure Regulators
ASCO Pneumatic Controls/
    Joucomatic Div.
The ARO Corp.
Automatic Switch Company
Barksdale Controls Div./
    Imo Industries Inc.
Deltrol Fluid Products
FESTO Corporation
Honeywell Inc., Skinner Valve Division
Legris Incorporated
Master Pneumatic-Detroit Inc.
Mecman Inc.
Miller Fluid Power
Parker Hannifin Corp.
Racine Fluid Power -
    Bosch Pneumatics
Rexroth Corp., Pneumatics Division
Ross Operating Valve Co.
Schrader Bellows
Snap-Tite Inc.
TAIYO AMERICA, INC.
Teknocraft, Inc.
Versa Products Co., Inc.
Vickers Incorporated
Wilkerson Corporation

Valves - Specialty
ASCO Pneumatic Controls/
    Joucomatic Div.
Anderson Copper and Brass Company
The ARO Corp.
Automatic Switch Company
Barksdale Controls Div. and
    Boston Gear Div./Imo Industries Inc.
Control Line Equipment, Inc.
Crouzet Corp.
Deltrol Fluid Products
FABCO-AIR INC.
FESTO Corporation
Fluid Product Sales Div. – Dana Corp.
Honeywell Inc., Skinner Valve Division
Kepner Products Company
Legris Incorporated
Lexair, Inc.
G. W. Lisk Company, Inc.
M-C Industries Inc.
Mecman Inc.
Miller Fluid Power
Moog Controls Inc.
Mosier Industries
Parker Hannifin Corp.
Racine Fluid Power -
    Bosch Pneumatics
Rexroth Corp., Pneumatics Division

Sources for Pneumatic Components

Ross Operating Valve Co.
Schrader Bellows
Snap-Tite Inc.
Teknocraft, Inc.
Telemecanique, Inc.
Valves Inc./
　Division of ACE Controls Inc.
Versa Products Co., Inc.
Vickers Incorporated
World Wide Fittings Corp.

OTHER

Air Compressors
Gast Manufacturing Corporation

Air Logic/Air Logic Controls
The ARO Corp.
Crouzet Corp.
FESTO Corporation
Miller Fluid Power
Mosier Industries
Rexroth Corp., Pneumatics Division
Teknocraft, Inc.
Telemecanique, Inc.

Air-Oil Systems
Boston Gear Div./Imo Industries Inc.
Control Line Equipment, Inc.
FABCO-AIR INC.
Galland Henning Nopak, Inc.
Hennells Inc., Div. of C. M. Smillie
Mico Inc.
Miller Fluid Power
Mosier Industries

Air Pressure Amplifiers
FABCO AIR INC.
Galland Henning Nopak, Inc.
Haskel, Inc.
Teknocraft, Inc.

Counters, Indicators, Gauges
The ARO Corp.
Control Line Equipment, Inc.
Crouzet Corp.
FESTO Corporation
Hedland Division,
　Racine Federated Inc.
Miller Fluid Power
Parker Hannifin Corp.
Racine Fluid Power -
　Bosch Pneumatics
Rexroth Corp., Pneumatics Division
Schrader Bellows
Teknocraft, Inc.
Telemecanique, Inc.

Mounting Components
B-D Cylinder Products, Inc.
Cylinder Components, Inc.
FESTO Corporation
PHD, Inc.
Parker Hannifin Corp.

Pneumatic Accessories
ASCO Pneumatic Controls/
　Joucomatic Div.
Almo Manifold & Tool Co.
Amflo Products
The ARO Corp.
Automatic Switch Company
Barksdale Controls Div./
　Imo Industries Inc.
BOBALEE Hydraulics
Control Line Equipment, Inc.
FABCO-AIR INC.
FESTO Corporation
Freudenberg-NOK General Partnership
Furon Co. – Synflex Division
Imperial Eastman, Imperial Division
Kilsby-Roberts
Mosier Industries
Nycoil Company
Quanex Lasalle Steel Co.
　Fluid Power Operations
Rexroth Corp., Pneumatics Division
J. H. Roberts Industries Inc.
W. S. Shamban & Co.
TAIYO AMERICA, INC.
Versa Products Co., Inc.

Pneumatic Feed Units
FESTO Corporation

Seals
Furon/Grover Piston Ring Division
Freudenberg-NOK General Partnership
Greene, Tweed & Co.
Hunger US Special Hydraulic
　Cylinders Corporation
Macrotech Fluid Sealing, Inc.
MERKEL Inc.
Miller Fluid Power
Minnesota Rubber
Parker Hannifin Corp.
W. S. Shamban & Co.

Suction Cups
Control Line Equipment, Inc.
FESTO Corporation
Mecman Inc.
Miller Fluid Power
PIAB Vacuum Products
Racine Fluid Power -
　Bosch Pneumatics
Teknocraft, Inc.

NFPA Member Companies that Manufacture Pneumatic Components

The NFPA, through its member companies, is dedicated to the advancement of fluid power technology. Through activities like the creation of this handbook, the NFPA fosters the refinement of pneumatics and hydraulics to ensure these technologies will serve as practical industrial power sources for today, and for many years to come.

AAA Products International
7114 Harry Hines Blvd.
Dallas, TX 75235
Telephone: (214) 357-3851
FAX: (214) 357-7223

ASCO Pneumatics Controls/ Joucomatic Division
8107-S Arrowridge Blvd.
Charlotte, NC 28217
Telephone: (704) 527-4622
FAX: (704) 527-4312

ACE Controls Inc.
23435 Industrial Park Drive
Farmington, MI 48332
Telephone: (313) 476-0213
FAX: (313-476-2470

Advance Automation
3526 N. Elston Ave.
Chicago, IL 60618
Telephone: (312) 539-7633
FAX: (312) 539-7299

Air-Dro, Div. of Decatur Cylinder Inc.
1112 Brooks St. S.E.
Decatur, AL 35602-1606
Telephone: (205) 350- 2603
FAX: (205) 351-1264

Almo Manifold & Tool Company
11444 Kaltz St.
Centerline, MI 48015
Telephone: (313) 756-0500
FAX: (313) 756-1820

American Couplings Company
3100 West Randolph St.
Bellwood, IL 60104-1992
Telephone: (708) 547-1000
FAX: (708) 547-0467

Amflo Products
Div. of Bridge Products, Inc.
1111 E. McFadden Ave.
Santa Ana, CA 92705
Telephone: (714) 547-9000
FAX: (714) 547-2167

Anderson Copper & Brass Co.
255 East Industry Ave.
Frankfort, IL 60423-0637
Telephone: (815) 469-2211
FAX: (815) 469-2265

The ARO Corp.
Ingersoll-Rand Company
One ARO Center
Bryan, OH 43506
Telephone: (419) 636-4242
FAX: (419) 636-2115

Atlas Cylinder Corp.
Parker Hannifin Corp.
29289 Airport Road
Eugene, OR 97402-0079
Telephone: (503) 689-9111
FAX: (503) 688-6771

Automatic Switch Company
Emerson Electric Company
Hanover Road
Florham Park, NJ 07932
Telephone: (201) 966-2000
FAX: (201) 966-2628

B-D Cylinder Products Inc.
4635 S. Harlem
Berwyn, IL 60402
Telephone: (708) 484-9300
FAX: (708) 484-9302

Barksdale Controls Division
Imo Industries Inc.
3211 Fruitland Ave.
Los Angeles, CA 90058-0843
Telephone: (213) 589-6181
FAX: (213) 589-3463

Bimba Manufacturing Company
P.O. Box 68
Monee, IL 60449
Telephone: (708) 534-8544
FAX: (708) 534-5767

BOBALEE Hydraulics
Division of GOMACO CORP.
137 N. East St.
Laurens, IA 50554
Telephone: (712) 845-4554
FAX: (712) 845-2503

Boston Gear Divison
Imo Industries Inc.
14 Hayward St.
Quincy, MA 02171
Telephone: (617) 328-3300
FAX: (617) 479-6238

T.J. Brooks Co.
2233 West Mill Road
Milwaukee, WI 53209
Telephone: (414) 352-4250
FAX: (414) 352-5763

CEC Instruments Division
Imo Industries Inc.
955 Overland Ct.
San Dimas, CA 91773-9013
Telephone: (714) 599-3200
FAX: (714) 394-4400

Chicago Fittings Corp.
18th Ave. & 21st St.
Broadview, IL 60153
Telephone: (708) 344-4214
FAX: (708) 344-6410

Control Line Equipment, Inc.
3827 Willow Ave.
Pittsburgh, PA 15234-1809
Telephone: (412) 531-5511
FAX: (412) 531-5508

Crouzet Corp.
2445 Midway Road
Carrollton, TX 75006
Telephone: (214) 250-1647
FAX: (214) 250-3865

Cylinder Components Inc.
P.O. Box 483
Franklin Park, IL 60131
Telephone: (708) 941-0450
FAX: (708) 678-1179

Dana Corp.
Fluid Product Sales Division
Caller #10016
Toledo, OH 43699
Telephone: (419) 891-2800
FAX: (419) 891-2816

NFPA Member Companies that Manufacture Pneumatic Components

Deltrol Fluid Products
Deltrol Corp.
3001 Grant Ave.
Bellwood, IL 60104
Telephone: (708) 547-0500
FAX: (708) 547-6881

FABCO-AIR INC.
3716 N.E. 49th Road
Gainesville, FL 32609
Telephone: (904) 373-3578
FAX: (904) 375-8024

Fenner Fluid Power
Div. of Fenner America, Inc.
5885 11th St.
Rockford, IL 61109-3699
Telephone: (815) 874-5556
FAX: (815) 874-7853

FESTO Corporation
395 Moreland Road
Hauppauge, NY 11788
Telephone: (516) 435-0800
FAX: (516) 435-8026

Flairline Division/ACE Controls Inc.
P.O. Box 439
Farmington, MI 48332-0439
Telephone: (313) 478-3330
FAX: (313) 476-2470

Freudenberg-NOK General Partnership
47690 E. Anchor Court
Plymouth, MI 48170
Telephone: (313) 451-0020
FAX: (313) 451-0125

Furon, Grover Piston Ring Div.
2759 South 28th St.
Milwaukee, WI 53215
Telephone: (414) 384-9472
FAX: (414) 384-0201

Furon Co. – Synflex Division
Main & Orchard Sts.
Mantua, OH 44255
Telephone: (216) 274-3171
FAX: (216) 274-0473

Galland Henning Nopak, Inc.
1025 South 40th St.
Milwaukee, WI 53215
Telephone: (414) 645-6000
FAX: (414) 645-6048

Gast Manufacturing Company
P.O. Box 97
Benton Harbor, MI 49002
Telephone: (616) 926-6171
FAX: (616) 926-0808

The Gates Rubber Company
990 South Broadway
P.O. Box 5887
Denver, CO 80217
Telephone: (303) 744-1911
FAX: (303) 744-4000

Greene, Tweed & Co. Inc.
Detwiler Road
P.O. Box 305
Kulpsville, PA 19443-0305
Telephone: (215) 256-9521
FAX: (215) 256-0189

Hanna Corporation
1765 N. Elston Ave.
Chicago, IL 60622
Telephone: (312) 384-7000
FAX: (312) 384-5224

Hansen Coupling Division
Tuthill Corp.
1000 West Bagley Road
Berea, OH 44017
Telephone: (216) 826-1115
FAX: (216) 826-0115 (Ex. & Mktg.)

Haskel, Inc.
100 East Graham Place
Burbank, CA 91502
Telephone: (818) 843-4000
FAX: (818) 841-4291

Hedland Division
Racine Federated Inc.
2200 South St.
Racine, WI 53404
Telephone: (414) 639-6770
FAX: (414) 639-2267

Hennells Inc.
Div. of C.M. Smillie Co.
1200 Woodward Heights Blvd.
Ferndale, MI 48220
Telephone: (313) 545-4120
FAX: (313) 545-1076

Hofmann Engineering Co. Inc.
280 Shore Dr.
Burr Ridge, IL 60521
Telephone: (708) 325-3744
FAX: (708) 325-3789

Honeywell Inc.
Skinner Valve Division
95 Edgewood Ave.
New Britain, CT 06051
Telephone: (203) 827-2300
FAX: (203) 827-2384

Horton Manufacturing Co.
1170 15th Ave. S.E.
Minneapolis, MN 55414
Telephone: (612) 331-5931
FAX: (612) 378-6496

Hunger US Special Hydraulic Cylinders Corp.
63 Dixie Highway
Rossford, OH 43460
Telephone: (419) 666-4510
FAX: (419) 666-9834

Hydro-Line Manufacturing Co.
Div. of M-C Industries Inc.
4950 Marlin Dr.
Rockford, IL 61130
Telephone: (815) 654-9050
FAX: (815) 654-3393

Imperial Eastman
Imperial Division
6300 W. Howard St.
Chicago, IL 60648
Telephone: (708) 967-4500
FAX: (708) 967-4616

Imo Industries Inc.
3450 Princeton Pike
Lawrenceville, NJ 08648
Telephone: (609) 896-7600
FAX: (609) 896-7688

Kepner Products Co.
995 North Ellsworth Ave.
Villa Park, IL 60181
Telephone: (708) 279-1550
FAX: (708) 279-9669

Kilsby-Roberts Company
3050 E. Birch St.
Brea, CA 92621
Telephone: (714) 579-8823
FAX: (714) 524-1072

Legris Incorporated
244 Paul Road
Rochester, NY 14624
Telephone: (716) 328-6250
FAX: (716) 328-6762

NFPA Member Companies that Manufacture Pneumatic Components

Lexair, Inc.
2025 Mercer Road
Lexington, KY 40511
Telephone: (606) 255-5001
FAX: (606) 255-6656

G. W. Lisk Company Inc.
2 South St.
Clifton Springs, NY 14432
Telephone: (315) 462-2611
FAX: (315) 462-7661

Lynair, Inc.
3515 Scheele Dr.
Jackson, MI 49204
Telephone: (517) 787-2240
FAX: (517) 787-4521

M-C Industries, Inc.
4251 Plymouth Road
Ann Arbor, MI 48105
Telephone: (313) 761-4400
FAX: (313) 761-4849

Macrotech Fluid Sealing, Inc.
Polyseal Division
1750 West 500 South
Salt Lake City, UT 84104
Telephone: (801) 973-9171
FAX: (801) 973-9188

Macrotech Fluid Sealing, Inc.
C.D.I. Division
8103 Rankin Road
Humble, TX 77396
Telephone: (713) 446-6662
FAX: (713) 446-7458

Mailhot Hydraulique (1988) Inc.
2721 Route 341 North
St. Jacques de Montcalm, Quebec
CANADA JOK 2RO
Telephone: (514) 861-9911
FAX: (514) 839-7419

Master Pneumatic-Detroit Inc.
6701 Eighteen Mile Road
Sterling Heights, MI 48078
Telephone: (313) 254-1000
FAX: (313) 254-6055

Mecman Inc.
1117 North Main St.
Lombard, IL 60148
Telephone: (708) 627-7200
FAX: (708) 627-7240

MERKEL Inc.
5375 Naiman Parkway
Cleveland, OH 44135
Telephone: (216) 248-2660
FAX: (216) 248-3142

Mico Inc.
1911 Lee Boulevard
North Mankato, MN 56001
Telephone: (507) 625-6426
FAX: (507) 625-3212

Micro-Precision Operations Inc.
525 Berne St.
Berne, IN 46711
Telephone: (219) 589-2136
FAX: (219) 589-8966

Miller Fluid Power Corp.
800 North York Road
Bensenville, IL 60106
Telephone: (312) 766-3400
FAX: (312) 766-3012

Milwaukee Cylinder
A Versa/Tek Company
5877 S. Pennsylvania Ave.
Cudahy, WI 53110
Telephone: (414) 769-9700
FAX: (414) 769-0157

Minnesota Rubber
3630 Wooddale Ave.
Minneapolis, MN 55416
Telephone: (612) 927-1400
FAX: (612) 927-1422

Moog Controls Inc.
Industrial Division
300 Jamison
East Aurora, NY 14052-3300
Telephone: (716) 655-3000
FAX: (716) 655-1803

Mosier Industries, Inc.
325 Carr Dr.
Brookville, OH 45309
Telephone: (513) 833-4033
FAX: (513) 833-4205

Nycoil Company
57 South Ave.
Fanwood, NJ 07023
Telephone: (201) 322-6644
FAX: (201) 322-9467

Origa Corporation
928 North Oaklawn Avenue
Elmhurst, IL 60126-1046
Telephone: (708) 832-4321
FAX: (708) 941-3424

Ortman Fluid Power
Coltec Industries Inc.
19-143rd St.
Hammond, IN 46327
Telephone: (219) 931-1710
FAX: (219) 931-2747

PHD Inc.
13 & Piper Dr., Baer Field
Fort Wayne, IN 46809
Telephone: (219) 747-6151
FAX: (219) 747-6754

Parker Hannifin Corp.
17325 Euclid Ave.
Cleveland, OH 44112-1290
Telephone: (216) 531-3000
FAX: (216) 531-3000

Peninsular, Inc.
27650 Groesbeck Highway
Roseville, MI 48066-2781
Telephone: (313) 775-7211
FAX: (313) 775-4545

Perfecting Services, Inc.
Div. of M-C Industries Inc.
332 Atando Ave.
Charlotte, NC 28206
Telephone: (704) 334-9175
FAX: (704) 334-9002

PIAB Vacuum Products
65 Sharp St.
Hingham, MA 02043
Telephone: (617) 337-6250
FAX: (617) 337-8028

Purolator Products Company
Facet Filter Products Division
8439 Triad Dr.
Greensboro, NC 27409-9621
Telephone: (919) 668-4444
FAX: (919) 668-4453

Quanex LaSalle Steel Co.
Fluid Power Operations
1045 East Main St.
Griffith, IN 46319
Telephone: (219) 853-6780
FAX: (219) 853-6788

Racine Fluid Power -
Bosch Pneumatics
7505 Durand Ave.
Racine, WI 53406
Telephone: (414) 554-7100
FAX: (414) 554-7117

NFPA Member Companies that Manufacture Pneumatic Components

Rexroth Corp., Pneumatics Division
1953 Mercer Road
Lexington, KY 40511
Telephone: (606) 254-8031
FAX: (606) 255-3503

J.H. Roberts Industries Inc.
3158 Des Plaines Ave. #231
Des Plaines, IL 60018
Telephone: (708) 699-0080
FAX: (708) 699-0082

Ross Operating Valve Company
1250 Kirts Blvd
Troy, MI 48007
Telephone: (313) 362-1250
FAX: (313) 362-3085

S-P Fluid Power
A Figgie International Co.
10711 North Second Street
Rockford, IL 61130
Telephone: (815) 633-5046
FAX: (815) 633-9427

Schrader Bellows
200 West Exchange St.
Akron, OH 44309
Telephone: (216) 375-5202
FAX: (216) 375-1355

W. S. Shamban & Company
Seals Division
2531 Bremer Dr.
Fort Wayne, IN 46803
Telephone: (219) 749-9631
FAX: (219) 749-0066

Sheffer Corporation
6990 Cornell Road
Cincinnati, OH 45242
Telephone: (513) 489-9770
FAX: (513) 489-3034

Sigma-Netics Inc.
One Washington Ave.
Fairfield, NJ 07006
Telephone: (201) 227-6372
FAX: (201) 882-0662

Snap-Tite Inc.
3250 West Lake Road
Erie, PA 16505
Telephone: (814) 833-6411
FAX: (814) 838-6382

TAIYO AMERICA, INC.
700 Frontier Way
Bensenville, IL 60106
Telephone: (708) 350-8810
FAX: (708) 350-8814

Teknocraft Inc.
1320 Clearmont St., Suite 108
Palm Bay, FL 32905
Telephone: (407) 729-9634
FAX: (407) 768-8732

Telemecanique, Inc.
2002 Bethel Road
Westminister, MD 21157
Telephone: (301) 876-2214
FAX: (301) 857-7577

Valves Inc./ACE Controls Inc.
P.O. Box 439
Farmington, MI 48332-0439
Telephone: (313) 474-5535
FAX: (313) 476-2470

Versa Products Co. Inc.
22 Spring Valley Road
Paramus, NJ 07652
Telephone: (201) 843-2400
FAX: (201) 843-2931

Vickers, Incorporated
A TRINOVA Company
5445 Corporate Drive
Troy, MI 48007-0302
Telephone: (313) 641-4200
FAX: (313) 641-4680

Wilkerson Corporation
1201 W. Mansfield Ave.
Englewood, CO 80110
Telephone: (303) 761-7601
FAX: (303) 781-8462

World Wide Fittings Corp.
6455 North Avondale Ave.
Chicago, IL 60631
Telephone: (312) 775-2121
FAX: (312) 775-1609

Index

Actuators	23, 26, 28, 40, 57
Air collets	22, 56
Air compressors	59
Air drills	24, 56
Air logic/air logic controls	59
Air motors	23, 56
Air-oil separators	59
Air-oil systems	59
Air pressure amplifiers	59
Air tools	24, 56
Air vises	25, 56
Assembly application	10
Automation	
Applications	6 - 17
Cost reduction	3, 4
Payback	7, 9, 11, 13, 15, 17
Step-by-step guide	18
Trouble-shooting guide	20
Boosters	26, 56
Broaching application	14
Component sources	56 - 59
Components	22 - 53
Cost reduction	3, 4
Chucks	22, 56
Clamping application	6
Clamps	25, 56
Connectors	35, 57
Counters, indicators, gauges	59
Cylinders	28, 56
Drilling application	12
Dryers	29, 56
Ejectors	46, 58
Electropneumatic converters	31, 56
Electropneumatic positioners	31, 56
FRL's	33, 56
Facing application	8
Filters	33, 56
Fittings	35, 57

Grippers	36, 57
Hose	35, 57
Index tables	37, 57
Information sources	54
Intensifiers	26, 56
Linear decelerators	43, 57
Loading application	16
Lubricators	33, 56
Mounting components	59
Mufflers	45, 57
Nut drivers	24, 56
PLC's	39, 57
Pneumatic accessories	59
Pneumatic feed units	59
Presses	38, 57
Programmable logic controllers	39, 57
Regulators	33, 56
Rotary actuators	40, 57
Screwdrivers	24, 56
Seals	59
Sensors	42, 57
Shock absorbers	43, 57
Silencers	45, 57
Suction cups	46, 59
Switches	42, 57
Tubing	35, 57
Vacuum ejectors	46, 58
Vacuum pumps	46, 58
Valves	
Check	51, 58
Directional control	47, 58
Flow control	48, 58
Impulse	53, 58
Pressure regulator	33, 50, 58
Quick exhaust	51, 58
Shuttle	51, 58
Specialty	51, 58
Time delay	53, 58